Eagan Press Handbook Series

Enzymes

Paul R. Mathewson

St. Paul, Minnesota, USA

Cover: Enzyme model courtesy of Novo Nordisk A/S; laboratory products and equipment courtesy of the Department of Food Science and Nutrition, University of Minnesota, St. Paul.

Library of Congress Catalog Card Number: 97-76811
International Standard Book Number: 0-913250-96-1

©1998 by the American Association of Cereal Chemists, Inc.

All rights reserved.
No part of this book may be reproduced in any form, including photocopy, microfilm, information storage and retrieval system, computer database or software, or by any means, including electronic or mechanical, without written permission from the publisher.

Reference in this publication to a trademark, proprietary product, or company name is intended for explicit description only and does not imply approval or recommendation of the product to the exclusion of others that may be suitable.

Printed in the United States of America on acid-free paper

American Association of Cereal Chemists
3340 Pilot Knob Road
St. Paul, Minnesota 55121-2097, USA

About the Eagan Press Handbook Series

The Eagan Press Handbook series was developed for food industry practitioners. It offers a practical approach to understanding the basics of food ingredients, applications, and processes—whether the reader is a research chemist wanting practical information compiled in a single source or a purchasing agent trying to understand product specifications. The handbook series is designed to reach a broad readership; the books are not limited to a single product category but rather serve professionals in all segments of the food processing industry and their allied suppliers.

In developing this series, Eagan Press recognized the need to fill the gap between the highly fragmented, theoretical, and often not readily available information in the scientific literature and the product-specific information available from suppliers. It enlisted experts in specific areas to contribute their expertise to the development and fruition of this series.

The content of the books has been prepared in a rigorous manner, including substantial peer review and editing, and is presented in a user friendly format with definitions of terms, examples, illustrations, and trouble-shooting tips. The result is a set of practical guides containing information useful to those involved in product development, production, testing, ingredient purchasing, engineering, and marketing aspects of the food industry.

Acknowledgment of Sponsors for *Enzymes*

Eagan Press would like to thank the following companies for their financial support of this handbook:

Amano Enzyme U.S.A. Co., Ltd.
Lombard, Illinois
800/446-7652

Danisco Ingredients USA, Inc.
New Century, Kansas
800/255-6837

Enzyme Development Corporation
New York, New York
212/736-1580

Eagan Press has designed this handbook series as practical guides serving the interests of the food industry as a whole rather than the individual interests of any single company. Nonetheless, corporate sponsorship has allowed these books to be more affordable for a wide audience.

Acknowledgments

Eagan Press thanks the following individuals for their contributions to the preparation of this book:

C. Peter Moodie, Enzyme Development Corporation, New York, NY

Clyde E. Stauffer, Technical Foods Consultants, Cincinnati, OH

Gary Yingling, McKenna & Cuneo, L.L.P., counsel for Enzyme Technical Association, Washington, DC

Contents

1. Basic Concepts • 1
What Is an Enzyme?: structure of proteins and enzymes
Enzyme Activity: basic chemical kinetics • basic enzyme kinetics
Orientation of the Substrate by the Enzyme: lock and key model • induced-fit model
Factors Affecting the Rate of Enzyme Reactions: effect of pH • effect of temperature • effect of substrate concentration
Enzyme Nomenclature

2. Production, Storage, and Handling • 13
Commercial Production of Enzymes: some safety and regulatory aspects
Storage of Enzymes: temperature and moisture levels • maintaining enzyme activity
Handling and Use of Enzymes: some practical examples • some safety precautions

3. Common Enzyme Reactions • 25
Enzyme-Substrate Interactions
Enzymes That Hydrolyze Carbohydrates: starch carbohydrases • cellulases • other carbohydrases
Enzymes That Hydrolyze Proteins: specificity • classification • action on protein
Enzymes Affecting Fats and Oils: lipase and other esterases • lipoxygenase
Other Enzymatic Reactions: oxidation and rancidity • enzymatic browning • enzymes involved in fruit ripening • other reactions

4. Analysis of Enzyme Activity • 41
Basic Principles
Types of Assay Methods: spectrophotometric tests • viscometric tests • pH • fluorescence • ELISA • comparison of test results
Specifications

5. Application of Enzymes to Baked Products • 49
Sources of Amylases
Bread: dough • product characteristics
Low-Moisture Products
Troubleshooting

6. Enzyme Applications for Beverages • 59

Beer and Ethanol Production: barley malting and enzyme activity • role of enzymes in the brewing process • low-calorie beers • distilled products and ethanol production

Wine Production

Fruit Juice Processing: noncitrus fruit juice • citrus fruit juice

Troubleshooting

7. Other Applications • 73

Commercial Sweetener Production

Dairy Application: cheesemaking • surface-ripened cheeses • enzyme-modified cheeses

Protein Modification

Endogenous Enzymes as Processing Indicators

8. Choosing Enzymes for Specific Applications • 81

An Approach to Solving Problems

New Opportunities for the Future

Appendix A. Amino Acid Side Chains • 89

Appendix B. Description of Selected Enzymes • 91

Appendix C. Major Biological Sources of Enzymes • 93

Appendix D. Specific Enzyme Assay Techniques • 97

Glossary • 103

Index • 107

Enzymes

Basic Concepts

What Is an Enzyme?

Enzymes are a type of *protein* present in, and essential to, all living things. They have a number of functions in the living cell, but ultimately, they help to convert food into energy and new material for the growth and repair of the organism in which they function. Enzymes act as biological catalysts—they increase the rate of chemical reactions without undergoing any permanent change themselves. They are not consumed in the reaction and can therefore continue to catalyze a reaction as long as the proper reactants, usually called substrates, are available.

STRUCTURE OF PROTEINS AND ENZYMES

Because the basic properties of enzymes are determined by their protein nature, it is useful to look at proteins in greater detail. First, all proteins are composed of *amino acids*. Each amino acid consists of an amino function ($-NH_2$) and a carboxylic acid function (COOH) attached to the same carbon atom. The general formula for an amino acid is:

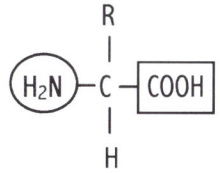

where the amino function is shown in the oval and the carboxylic acid function is shown in the rectangle. The differences in the individual amino acids (there are about 20 that occur in nature) are determined by the variations in the "R" group, sometimes called the "side chain." The amino acid side chains may contain additional acid groups or amino groups as well as other functional groups. Appendix A shows all the common amino acids and their side chains.

The amino function and the carboxylic acid function are chemically reactive. Two amino acids can combine by forming a peptide bond (also referred to as an amide bond) between them. The compound formed by connecting two amino acids through a peptide bond is called a dipeptide. This reaction, which also yields a molecule of water, is illustrated in Figure 1-1, in which the nitrogen atom of the

---

In This Chapter:

What Is an Enzyme?
  Structure of Proteins and Enzymes

Enzyme Activity
  Basic Chemical Kinetics
  Basic Enzyme Kinetics

Orientation of the Substrate by the Enzyme
  Lock and Key Model
  Induced-Fit Model

Factors Affecting the Rate of Enzyme Reactions
  Effect of pH
  Effect of Temperature
  Effect of Substrate Concentration

Enzyme Nomenclature

Proteins—Polymers composed of amino acids. Enzymes are proteins.

Amino acids—A group of organic compounds having the general formula $NH_4C_2O_2R$. Structurally, each is a carboxylic acid with an amino group attached to the α-carbon atom. R represents a functional group peculiar to each amino acid.

amino group is blue, the carbon atoms are black, and the oxygen atoms of the acid group are red. Hydrogen atoms are attached to the nitrogen, carbon, and oxygen atoms as shown in the general formula for an amino acid, but they are omitted here for clarity.

The large oval portion in Figure 1-1 represents the side chain of the amino acid. When the reaction just illustrated occurs many times, the result is a chain of amino acids. A protein is the large molecule (a

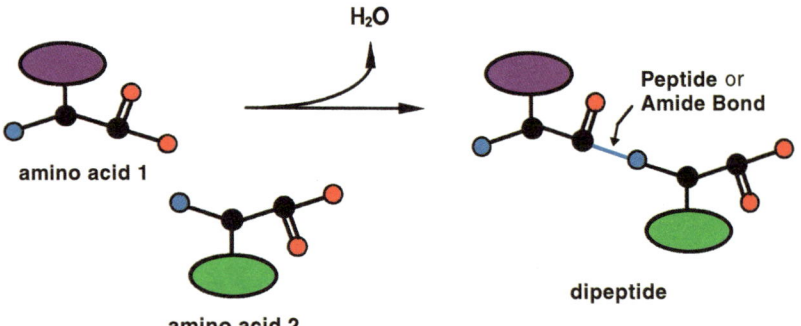

Fig. 1-1. Formation of a dipeptide from two amino acids joined with a peptide bond. Blue = nitrogen atom, black = carbon atom, red = oxygen atom.

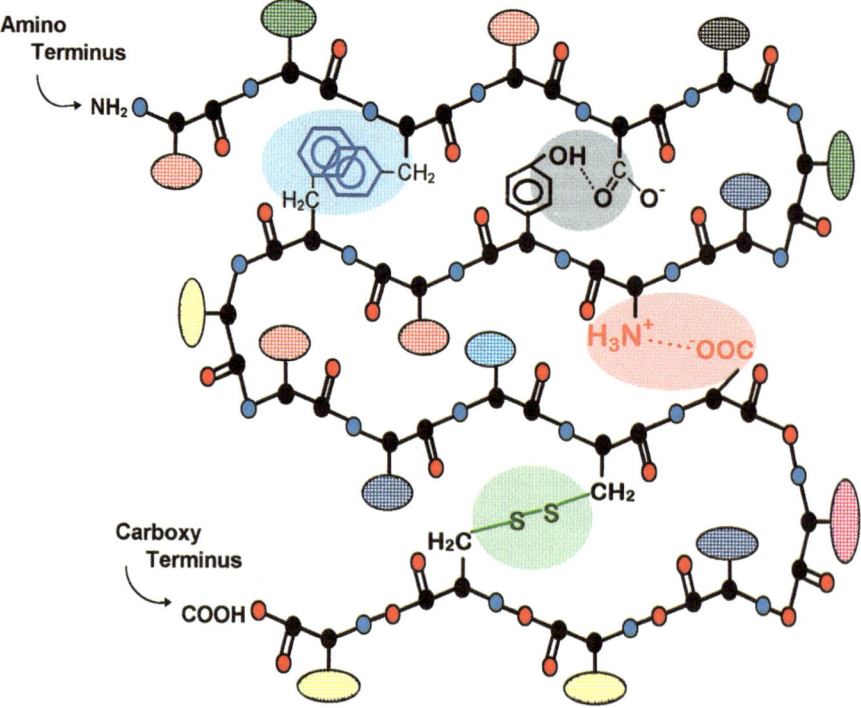

Primary structure—The sequence of amino acids making up a protein, from amino terminus to carboxy terminus.

Secondary structure—Protein configuration stabilized by hydrogen bonds not involving side chains; characterized as an α-helix or a β-pleated sheet.

Fig. 1-2. A protein polymer composed of a linear combination of amino acids. Forces holding the protein in proper conformation include hydrophobic interaction (blue), disulfide bonds (green), hydrogen bonds (gray), and electrostatic interaction (red).

polymer) formed by the combination of many smaller molecules (amino acids) (Fig. 1-2). The different functional side chains on each amino acid making up the protein contribute to the overall properties of any given protein.

As a protein grows in size, its structure becomes more complex. The *primary structure* is the sequence of amino acids along the polymer chain from the amino end of the protein to the carboxylic acid end. Even a small change in this sequence, such as the change of one amino acid to a different amino acid, can have a marked effect on the functional characteristics of the protein.

A protein usually contains about 150–6,000 amino acids. That translates to molecular weights from about 15,000 to 600,000. When neighboring sections of the protein fold in specific ways, again depending on the sequence of amino acids, another level of structure is formed called the *secondary structure*. This may be defined as regular, repeating patterns formed by the peptide backbone of part of the polypeptide chain, usually stabilized by *hydrogen bonds*. These structures are usually either in the form of an α-helix or a pleated sheet (Fig. 1-3).

The next level of structure, called the *tertiary structure*, is described by the overall spherical shape achieved by the folding of the entire protein molecule. Relatively weak bonds between amino acids that are in close proximity (such as electrostatic forces between charged amino acids and hydrophobic interactions between nonpolar amino acids) help stabilize this level of protein structure. In addition, fairly strong bonds can be formed. For example, a *covalent bond* can be formed between two cysteine amino acids in which their sulfur atoms form a *disulfide bond*. These kinds of inter-amino-acid bonds play an important role in determining the overall conformation of a protein. The bonds may be formed by amino acids that are widely separated in terms of the primary structure but are in close proximity due to the twisting and bending of the polypeptide chain. These interactions are illustrated in Figure 1-2. The disulfide bonds are shown in green; the electrostatic attraction between the NH_3^+ of a basic amino acid and the COO^- of an acidic amino acid is shown in red; the hydrophobic interaction between two nonpolar residues such as phenylala-

Hydrogen bonds—Relatively weak interactions between a hydrogen and an electronegative atom such as oxygen or nitrogen.

Tertiary structure—Overall three-dimensional folding of a protein chain.

Covalent bond—A chemical bond formed when two atoms share an electron.

Disulfide bond—A sulfur-sulfur bond formed between two cysteine amino acid residues, which helps to stabilize the folded structure of proteins.

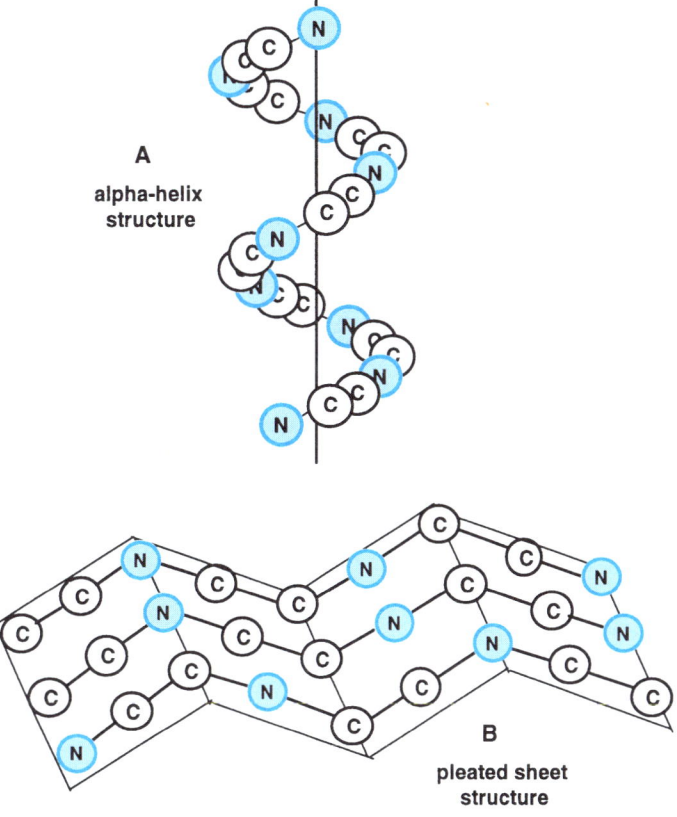

Fig. 1-3. Secondary structure of a protein, showing two arrangements of carbon and nitrogen atoms: α-helix (A), pleated sheet (B).

nine is in the blue field; and the hydrogen bond is illustrated in the gray field.

These levels of structure are very important because they determine how proteins, and of course enzymes, function. The tertiary structure results in a configuration that may be thought of as a ball of yarn. That is, the chain of amino acids, represented by the yarn, flows in and around, back on itself, over and over and the result is a basically spherical form like a ball.

The last level of structure in proteins occurs when several polypeptide chains, sometimes identical, sometimes not, interact to form one large protein. The complete three-dimensional structure, including the interactions between the component polypeptide chains, is referred to as the protein's *quaternary structure*. These proteins, too, have an essentially overall spherical structure. Most protein molecules, whether large or small, are more or less spherical when in a water solution.

Enzyme Activity

Enzymes are proteins with the special ability to catalyze, or speed up, the rate of chemical reactions under the rather mild conditions found in living organisms. For example, sucrose can be *hydrolyzed*, or split into its component sugars, glucose and fructose, by being heated in acid. This requires relatively harsh conditions that would not be suitable for sucrose conversion in most living things. In living organisms, the conversion of sucrose is brought about by an enzyme called *invertase*. Not only can the enzymatic reaction proceed under mild conditions, the rate of the enzymatic reaction is over 50,000,000,000 times faster than that of the acid-heat reaction!

BASIC CHEMICAL KINETICS

Chemical reactions can take place with or without enzymes. In the absence of an enzyme, chemical reactions occur, but they depend on a number of factors. First, the reactants must find and meet each other. This is normally the result of random collisions of reactant molecules in some medium such as water. Second, just any collision is not sufficient to result in the formation of a product. The molecules must collide with sufficient energy and in the proper orientation so that they form a complex, sometimes called a reactive intermediate. Often when collisions occur the orientation of the molecules is not correct and there is not enough energy involved to rearrange the bonds, so the reactive intermediate cannot develop and no new products are formed. The rearrangement of bonds requires, among other things, the input of energy. This *energy of activation* represents an "energy barrier." The reactant molecules must collide with at least this amount of energy to overcome the barrier so that the intermediate can be formed and the reaction can proceed to the formation of products.

Quaternary structure—The overall spatial structure of a protein containing more than one polypeptide chain.

Hydrolysis—The breaking of a chemical bond(s) through a water-mediated decomposition mechanism.

Invertase—Enzyme catalyzing conversion of sucrose to component sugars, glucose and fructose.

Energy of activation—The minimum energy required to convert a normal reactant molecule into a reactive intermediate.

BASIC ENZYME KINETICS

The kinetics of enzyme reactions are similar to those for nonenzymatic reactions except that enzymes have a unique way of accelerating the reaction. They act as catalysts by bringing the reactants together so as to promote the chemical reaction. First there is the rapid combination of the enzyme (E) with the reactant molecule or substrate (S) to form a reactive intermediate, in this case, the enzyme-substrate complex (ES). This complex then breaks down into products (P) and releases the enzyme, unchanged, to combine with another substrate molecule. This is represented in the following general equation for a one-substrate enzyme-catalyzed reaction.

$$E + S \underset{k_{-1}}{\overset{k_1}{\rightleftarrows}} ES \underset{k_{-2}}{\overset{k_2}{\rightleftarrows}} E + P \qquad (1)$$

As indicated by the arrows, the formation of the enzyme-substrate complex and the breakdown of ES to product can both be reversible reactions. The coefficients k_1, k_{-1}, k_2, and k_{-2} represent the rate constants for each reaction and depend on the specific conditions. The formation of the ES complex is usually rapid, whereas the breakdown to products is a slower, rate-limiting reaction that normally proceeds in an energetically preferred direction (i.e., k_2 is normally much greater than k_{-2}). However, under certain conditions, an enzyme-catalyzed reaction can proceed in either direction. For example, some enzyme-catalyzed reactions that proceed in a forward direction in aqueous media can be made to proceed in the reverse direction in a nonaqueous media.

The rate, v, at which product is formed is proportional to the concentration of the ES complex, according to the equation:

$$v = k_2 (ES) \qquad (2)$$

Equations 1 and 2 provide a basis for calculation of reaction rates, substrate-binding energy, and many other important characteristics of enzymatically catalyzed reactions.

In a simple visual form, equation 1 may be represented by Figure 1-4. This figure shows that the substrate, with its own special shape, fits with the particular enzyme catalyzing this reaction in just the right orientation. This leads to a decrease in the energy of activation, thus allowing the reaction to proceed more easily to form products and free enzyme.

Figure 1-5 shows a schematic representation of this process. The blue curve represents the energy profile for a nonenzymatic reaction such as the acid hydrolysis of sucrose; the red curve represents the energy profile for the analogous enzymatically catalyzed reaction, e.g., hydrolysis of sucrose by invertase. The energy required to drive the nonenzymatic reaction is significantly greater than that required for the enzymatic reaction. At the bottom left under the curve, the reac-

Active site—The region on the surface of an enzyme where catalytic activity occurs.

tants (enzyme and substrate) are shown coming together. The area below the curves contains the scheme shown in Figure 1-4, to illustrate how the reaction proceeds.

One important concept in enzyme kinetics is reaction order, defined as zero-order, first-order, second-order, or third-order based on the rate of reaction under defined conditions where only the concentration of reactants varies and the enzyme concentration is held constant. In a zero-order reaction, the rate of reaction is independent of reactant (substrate) concentration. In first-, second-, and third-order reactions, the rate is dependent on one, two, or three reactants, respectively. It is important to keep this in mind when assaying enzyme activity. The substrate concentration must be kept very high relative to the enzyme concentration so that the reaction is zero-order with respect to substrate. Under such conditions (at constant pH, temperature, etc.), the reaction is independent of substrate concentration and depends on enzyme concentration only.

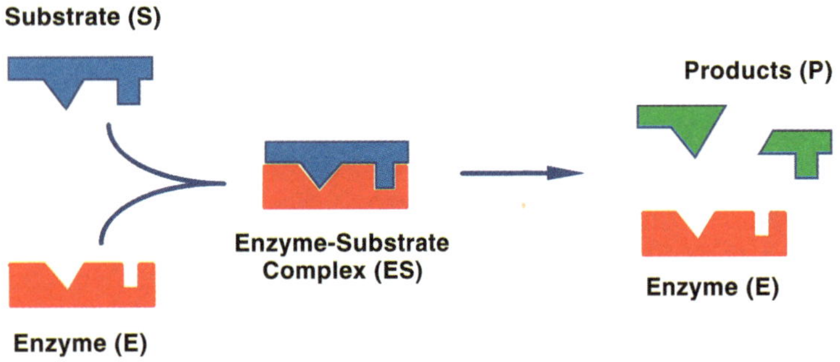

Orientation of the Substrate by the Enzyme

An enzyme can speed the rate of reaction by guiding the reactants (the reactant molecule as well as another molecule such as water or a hydrogen ion) together so that the interaction is not just random but is directed in such a way as to facilitate the formation of the reactive intermediate.

As noted earlier, proteins in water solution are basically spherical. While that describes the general shape of the molecule, the surface is not really smooth. The surface is actually quite irregular, having high spots as well as crevasses. Depending on how the folding of the protein chain occurred, certain amino acid side chains are exposed at precise locations on the surface of the protein sphere and others are buried in the middle. In an enzyme, specific amino acids are exposed at precise locations on the molecule's surface. These are *active sites*, where the

Fig. 1-4. Schematic representation of an enzymatic reaction. Blue = substrate, red = enzyme.

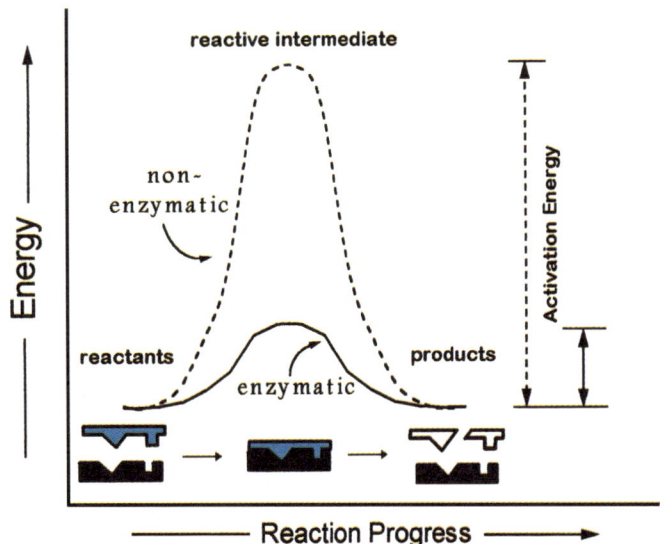

Fig. 1-5. Comparison of enzymatic and nonenzymatic reaction kinetics, showing the differences in activation energies for each as the reaction proceeds from reactants to products. Blue = substrate, black = enzyme.

substrate binds to the enzyme and the actual chemical reaction takes place. The amino acids exposed in the active site are specifically suited to attract and capture a reactant molecule in just the right orientation so that a chemical reaction can occur quickly and efficiently. Because the active site usually will allow only one substrate molecule to fit, the reaction also is very specific to that substrate.

Lock and key model—A theory to explain enzyme specificity in which a substantially rigid active site is likened to a lock and the substrate to a key that fits the lock.

LOCK AND KEY MODEL

A number of theories have been offered to explain exactly how the enzyme is able to catalyze a specific reaction. One involves a mechanism called the *lock and key* relationship between the reactant and the enzyme (1). This is illustrated in Figure 1-6. The figure is similar to Figure 1-4 but is more detailed in that it shows more of the chemical interactions involved in the reaction. In the lock and key theory, the active site of the enzyme represents a rigid lock, and the substrate is the rigid key that fits it. It may be that some other molecules are close enough in shape to be able to enter and bind to the active site, but that does not mean that a reaction will take place.

An enzymatic reaction actually involves two steps: a binding step in which the substrate is attracted to and captured by the enzyme, followed by the actual reaction step that forms the products. In Figure 1-6, the active site of the enzyme is the darker green area. This is the location on the surface of the enzyme containing the specific amino acids required for the chemical reaction. Some of the amino acids in the active site (blue dots) are required to attract and hold the substrate to the enzyme in a specific orientation. Other amino acids in the active site, shown as red dots, are involved in the actual chemical reaction leading to the breakdown of the substrate to product.

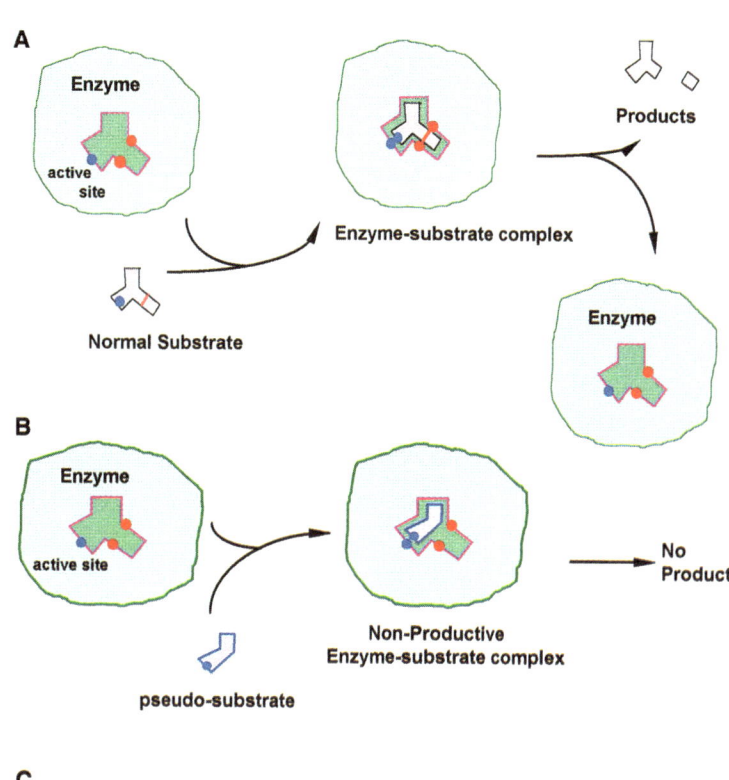

Fig. 1-6. Schematic representation of enzymatic reactions involving the binding and catalysis steps. A, a normal interaction between the enzyme and its proper substrate, leading to formation of products. B and C, interactions with similar, but incorrect, pseudosubstrates that fail to result in either catalysis (B) or binding (C). Light green area = enzyme, dark green area = active site, blue dot = amino acids that bind substrate, red dots = amino acids involved in the reaction, red line = reaction, pink areas = substrate, pseudosubstrate, or product.

In Figure 1-6A, a proper substrate molecule collides with and is bound to the amino acids in the active site, both at the blue binding site as well as the red catalytic site. The blue dot on the substrate represents an amino acid that is complementary to the amino acid in the active site of the enzyme. The red line on the substrate represents amino acids that are complementary to the amino acids in the enzyme's active site that are required to produce the reaction. When everything lines up just right, the reaction proceeds as it should, resulting in the rapid formation of product. Several other situations are possible, however.

In Figure 1-6B, a substrate molecule with an only partially correct shape, sometimes called a pseudosubstrate, enters the active site and may be able to bind, but no interaction with the catalytic amino acids can take place and no products are generated. These molecules may remain in the active site, preventing other correct substrate molecules from binding and forming products. When this happens, the enzyme is said to be inhibited.

Figure 1-6C shows a similar situation with a different pseudosubstrate. This one is not complementary to the binding site and will not be held in place properly. No reaction will result. Again, this situation may lead to inhibition of the enzyme, preventing or slowing further enzymatic activity.

INDUCED-FIT MODEL

Based on more recent investigations, a second theory has emerged that is referred to as the *induced-fit model* (2). This model of enzyme reaction is similar to the lock and key theory except that in the induced-fit model, a more flexible active site is partially prepared for the substrate. After the substrate interacts with the enzyme in a specific way, the enzyme is caused, or induced, to change its shape so that it fits exactly around the substrate molecule, thus leading to catalysis and the formation of products. There is abundant experimental evidence for both the lock and key and the induced-fit models. In fact, almost all enzymatic reactions studied so far can be explained using these two models for enzyme specificity and catalysis.

Factors Affecting the Rate of Enzyme Reactions

The rate at which enzyme reactions proceed is dependent on many factors, but among the most important are the pH of the reaction medium (usually an aqueous solution), the temperature at which the reaction occurs, and the amount of substrate available to react compared with the amount of enzyme present.

EFFECT OF pH

The forces that hold a protein chain in its particular shape result from a number of interactions, involving both the peptide backbone

Induced-fit model—A theory to explain enzyme specificity in which a flexible active site is induced, by a substrate, to alter its conformation to an orientation properly fitting the substrate's geometry.

(α-helix, pleated sheet) and the side chains (the R-groups) of the amino acids making up that protein. The side chains of the amino acid can be charged, either positively or negatively depending on the pH of the solution. At a neutral pH (7.0), most proteins have both positive and negative charges available along the amino acid chain. The opposite charges attract each other, while like charges repel each other. Although the force is not very strong, these repulsion/attraction forces play a significant role in maintaining the overall three-dimensional (tertiary) structure of the protein, which is so important to its functionality.

The ability of the amino acids at the active site of an enzyme to interact with the substrate depends on their electrostatic state, that is, whether they are properly charged or uncharged, as well as their spatial orientation. If the pH is not correct, the charge on one or all of the required amino acids is such that the substrate can neither bind nor react to produce product. Most enzymes have a rather narrow range in which they can catalyze a reaction efficiently. The majority work in the pH range of 6.0–8.0, although there are exceptions to this. For example, *pepsin*, the enzyme in your stomach that helps to digest protein, works best at a pH of about 2.0, which is very acidic. *Lipoxygenase* enzymes from soybeans work best around pH 9.0. In general, commercial enzymes prefer a pH around neutral. If the pH of the aqueous medium remains within about 1 pH unit of the optimal pH, the enzyme continues to function reasonably well but is less efficient and has a slower rate of reaction than at optimum pH. Therefore, for the most efficiency and lowest cost, one should always try to use the enzyme at its optimal pH. Figure 1-7 shows enzyme activity as a function of pH. The actual pH optimum should be determined experimentally for each enzyme of interest. The determination must take into account other operational variables such as the temperature and specific substrate used under the same conditions as the particular application. The specific chemical buffer used to maintain the pH can also have an effect on enzyme activity.

Care must be taken to prevent the pH from straying too far from the optimal value, or major problems can develop. If the pH is far below or above the optimal pH, the electrostatic forces holding the amino acid chain in its proper conformation may be altered. At low pH, the amino groups tend to be positively charged (NH_3^+), while the acid groups are neutral (COOH). The positive charges repel each other, driving the components of the enzyme apart. This results in alteration of the tertiary and secondary structure and, consequently, gross changes in the active site.

Pepsin—An acid protease, usually derived from the bovine or porcine digestive tract and useful in the dairy industry.

Lipoxygenase—An enzyme that catalyzes the oxidation of unsaturated fatty acids containing a *cis-cis* penta-1,4-diene unit to the corresponding monohydroperoxide.

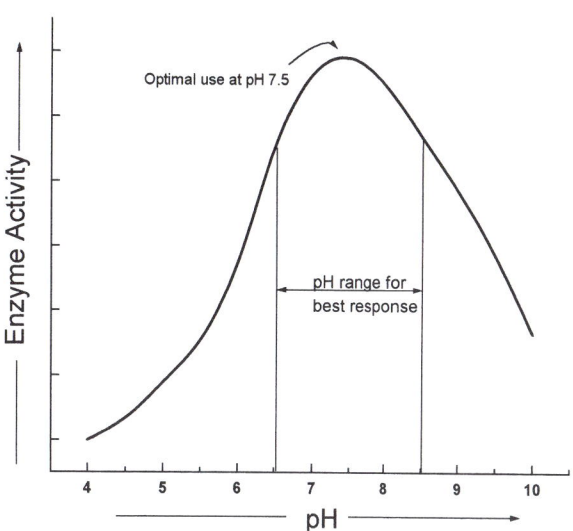

Fig. 1-7. Relationship between pH and activity for a typical commercial enzyme.

Denaturation—The loss of normal spatial arrangements in a polypeptide chain.

Renaturation—The reestablishment of the normal spatial arrangements in a polypeptide chain.

Ultimately the whole protein (enzyme) chain may unfold (Fig. 1-8). When this happens, the enzyme is said to be *denatured* and is usually permanently inactivated. Sometimes, if the environment of the enzyme is returned to more moderate conditions (less extreme pH), the enzyme may refold to its normal configuration and activity. If this occurs, the enzyme has undergone a spontaneous refolding called *renaturation*. While this does occur with a small number of proteins, it should not be considered a likely event with most commercial enzymes.

EFFECT OF TEMPERATURE

The effect of temperature is similar to the effect of pH in that some heat can help an enzyme but too much can permanently inactivate it. Putting more heat into the system helps to overcome the energy barrier. This occurs for two reasons. First, the warmer the solution, the faster the molecules move. Thus, the reactants collide more often and with greater energy. In general, chemical reactions, including enzymatic reactions, obey what is known as the Q_{10} *rule* (Q-ten rule), which states that the rate of reaction increases by a factor of about two for every 10 degree C increase in temperature. This is true for many enzymatic reactions, including those using commercial enzyme preparations for the food industry, until the temperature reaches about 60–70°C. At about this temperature range, the energy introduced to the system begins to overcome the energy of the attractive forces holding the enzyme in its required conformation. After a sufficient period of time, the enzyme will unfold (denature) and be inactivated, usually permanently. The higher the temperature, the more quickly denaturation takes place. As with the pH example, there are exceptions to this general rule. Some enzymes derived from a specific bacterium are able to function in boiling water. Even in frozen foods, enzymes can continue to function, albeit at a much slower rate. However, for the majority of commercial enzymes, the optimal temperature range is usually about 40–60°C. The activity profile for a typical enzyme as a function of temperature is shown in Figure 1-9.

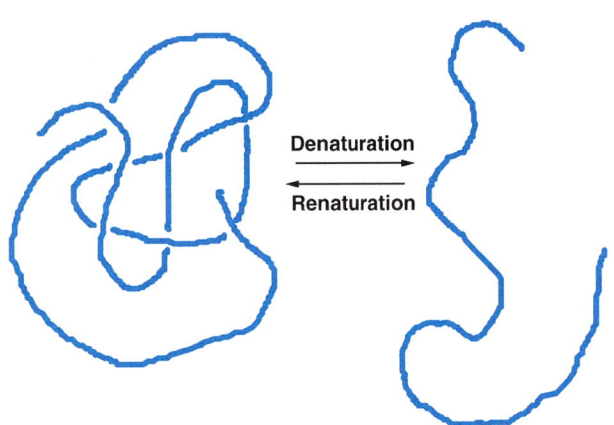

Fig. 1-8. Loss of three-dimensional structure (secondary and tertiary) of a protein as a result of denaturation. This process may be reversible in some instances.

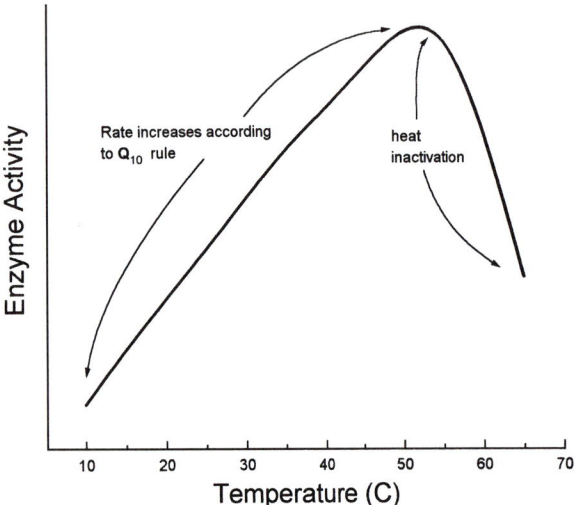

Fig. 1-9. Relationship between temperature and the activity of a typical commercial enzyme.

EFFECT OF SUBSTRATE CONCENTRATION

The last factor to be considered here is the concentration of substrate relative to the concentration of enzyme. The enzyme operates most efficiently when an excess of substrate is available. If the substrate concentration is low compared to that of the enzyme, successful collisions are few. When the substrate concentration is very high compared to the enzyme concentration, successful collisions occur very rapidly, ensuring that most of the enzyme is bound in an enzyme-substrate complex. Product is then produced at the maximum rate. In Figure 1-10, the curve shows how enzyme activity responds to substrate concentration. Actually, the important relationship is the ratio of substrate to enzyme. When this is low (in the area at the left of the graph), relatively little product is produced. When the ratio of substrate to enzyme is very high (in the area at the top of the graph), production of product is very rapid and the enzyme activity is at its maximum. This is a very important consideration in the design of analytical testing methods for enzyme activity.

Enzyme Nomenclature

In general, enzymes are named by taking the root of the word describing the substrate acted upon by the enzyme and adding -*ase*. For example, an enzyme that breaks down protein is called a *proteinase* or *protease*. Likewise, *pectinase* breaks down pectin. An enzyme that breaks down starch is called an *amylase*. Unfortunately, many enzymes were discovered and named before the adoption of this convention, and those names are still in use today. Some examples are *trypsin, papain, bromelain,* invertase, and *diastase*. To accommodate the naming of all the specific types of enzymes found in nature, rules for identifying enzymes have been devised by a committee made up of members of the International Union of Pure and Applied Chemistry and the International Union of Biochemistry. They use an identification system in which each enzyme is distinguished with a unique set of four numbers. The first number denotes the class of chemical reaction. Each succeeding number further identifies the specific reaction catalyzed by that enzyme by describing a subclass of the reaction identified by the previous digit. This results in a number called the EC (Enzyme Commission) number.

Q_{10} rule—Rule stating that a chemical reaction doubles in rate for every 10 degree C increase in temperature.

Protease (proteinase)—Enzyme that hydrolyzes proteins to peptides and/or amino acids.

Pectinase—Enzyme that degrades pectic substances.

Amylase—Enzyme that hydrolyzes starch to dextrins and/or sugars.

Trypsin—Proteolytic enzyme having limited specificity, derived from animal gut.

Papain—Proteolytic enzyme with broad specificity, derived from papaya.

Bromelain—Enzyme similar to papain derived from pineapple.

Diastase—An early name for the α-amylase enzyme.

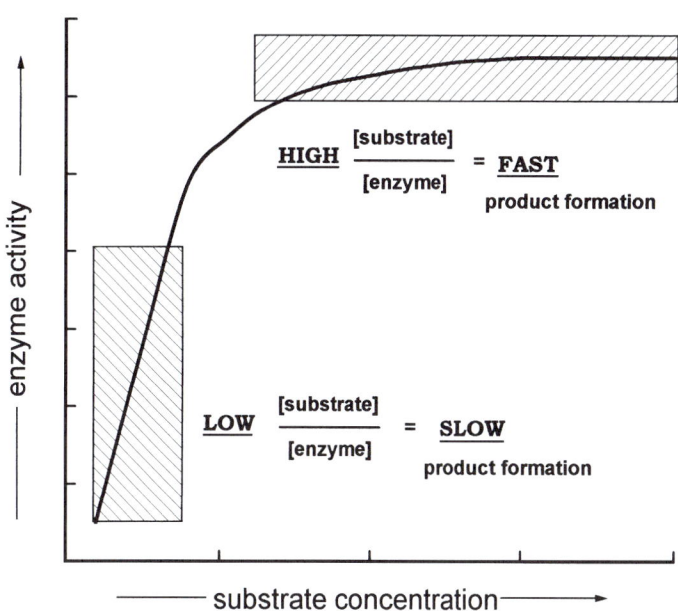

Fig. 1-10. Effect of substrate concentration on the rate of an enzymatic reaction.

For example, α-amylase breaks down starch. Its EC number is 3.2.1.1. The first number represents the class of chemical reaction. Those reactions are codified as follows:

1. Oxidoreductases—enzymes catalyzing oxidation/reduction reactions.
2. Transferases—enzymes catalyzing group transfer reactions.
3. Hydrolases—enzymes catalyzing hydrolytic reactions.
4. Lyases—enzymes catalyzing addition of groups to double bonds or vice-versa.
5. Isomerases—enzymes catalyzing group rearrangements.
6. Ligases—enzymes catalyzing combination of two molecules with involvement of adenosine 5′-triphosphate (ATP).

For α-amylase, The first number, 3, identifies the enzyme mechanism as belonging to the class of enzymes called hydrolases. The second number, 2, indicates the type of bond hydrolyzed in the hydrolase reaction. It is a glycosidic bond linking two sugar molecules, as shown in Table 1-1.

The third digit, 1, further defines the type of glycosidic bond hydrolyzed by α-amylase as an O-glycosyl bond between two sugar molecules (Table 1-2).

The fourth digit represents the serial number of the enzyme in that particular series. α-Amylase is the first in that series, so its last number is 1. There are many others in this series, the last one (so far) being glucan 1,4-α-maltohexaoside, with an EC number of 3.2.1.98. A list of important food enzymes, their EC numbers, and properties is shown in Appendix B. The listing is not complete but is meant to show the wide diversity of enzymes and how the numbering system applies to them.

TABLE 1-1. Second Digit in Enzyme Commission Number

Second Digit for Hydrolases	Bond Hydrolyzed
1	Ester
2	Glycosidic (links carbohydrate units)
3	Ether
4	Peptide
5	Carbon-nitrogen (other than peptide)
⋮	⋮
11	Carbon-phosphorous bond

TABLE 1-2. Third Digit in Enzyme Commission Number

Third Digit for Glucosidic Hydrolase	Bond Description
1	O-glycosyl compound
2	N-glycosyl compound
3	S-glycosyl compound

References

1. Fischer, E. 1894. Berichte der Deutschen Chemischen Gesellschaft 27:2985.
2. Koshland, D. E., Nemethy, G, and Filmer, D. 1966. Biochemistry 5:365.

Supplemental Reading

1. Whitaker, J. 1972. *Principles of Enzymology for the Food Sciences.* Marcel Dekker, New York.
2. Belitz, H. D., and Grosch, W. 1987. *Food Chemistry*. Springer-Verlag, New York.
3. Palmer, T. 1985. *Understanding Enzymes.* Ellis Horwood, London.
4. Berk, Z. 1976. *Braverman's Introduction to The Biochemistry of Foods*. Elsevier Scientific Publishing Co., London.
5. Fennema, O. W., Ed. 1996. *Food Chemistry.* Marcel Dekker, New York.
6. Alais, C., and Linden, G. 1991. *Food Biochemistry.* Ellis Horwood, London.
7. Godfrey, T., and Reichelt, J. 1983. *Industrial Enzymology: The Application of Enzymes in Industry*. The Nature Press, New York.

CHAPTER 2

Production, Storage, and Handling

The commercial production of enzymes is a relatively new industry, having started on a large scale in the 20th century. Many fundamental questions about the nature of enzyme activity had to be answered before commercial production could be pursued.

In the early 1800s, processes such as fermentation were recognized but poorly understood. Scientists of the time debated whether the process of fermentation was a simple chemical process or whether a living organism (yeast) had to be present. The debate was not settled until 1897, when it was demonstrated that a cell-free yeast extract could convert glucose to alcohol and carbon dioxide even though no viable, living yeast cells were present. This led to the realization that fermentation was not the result of the living yeast itself, but rather was attributable to the enzyme components of the yeast cell. The term *enzyme* was introduced by William Kuhne to describe these extracts from living organisms. The word comes from the Greek words *en*, meaning "in," and *zyme*, meaning "yeast" or "leaven."

In This Chapter:

Commercial Production of Enzymes
  Some Safety and Regulatory Aspects

Storage of Enzymes
  Temperature and Moisture Levels
  Maintaining Enzyme Activity

Handling and Use of Enzymes
  Some Practical Examples
  Some Safety Precautions

Commercial Production of Enzymes

Early industrial enzymes were derived from plants and animals. The plant or muscle material was chopped and mascerated in an appropriate liquid and then stirred to extract the soluble enzymes. The solid material was then removed by filtration, leaving the liquid containing the enzymatic activity. This could be further concentrated by evaporation. While this production system worked relatively well for the time, it was labor-intensive, produced varying amounts of enzyme, and was difficult to control. In addition, the growing awareness of the applicability of enzymes to many industrial processes increased the demand for them, and producing the necessary volume of enzyme material became difficult. As a result, microbial fermentation has largely replaced plant and animal extracts as the major source of industrial enzymes.

The production of enzymes from microbial fermentation is a more efficient method of manufacture, but it is a complicated process that is constantly evolving. The microbial sources may be from bacteria, fungi (molds), or yeasts. Some common microorganisms used to produce enzymes for food use are the fungi *Aspergillus niger* and *A. oryzae*, the bacteria *Bacillus subtilis* and *Streptomyces griseus,* and the yeasts

Kluyveromyces fragillis and *Saccharomyces cerevisae*. Great care is taken to maintain these organisms in a pure-culture state to ensure that no cross-contamination takes place.

In most cases, the organisms are grown in large tanks under carefully controlled conditions. Changes in the conditions of fermentation can alter the amount, and in some cases, even the type of enzyme produced. One way to increase the efficiency of the operation is to increase the propagation of the microbial cells under ideal conditions. The amount of enzyme produced increases in proportion to the increased number of cells. Alternatively, the media in which the fermentation takes place is formulated so as to maximize the rate of enzyme production from each cell. In many instances, a combination of both approaches is used.

The result of the fermentation is a liquid broth, which is concentrated by evaporation or *ultrafiltration*. Glucose isomerase is an exception to the rule. This enzyme is intra- rather than extracellular. Whole cells are filtered and then immobilized on frozen or other media. This is the only large-scale industrial application of whole-cell enzymes rather than pure protein extracts.

In most cases, the desired enzyme is excreted from the cell into the fermentation media, but in some instances, the enzyme remains in the microorganism. When this is the case, the cells are *lysed*. In some cases, depending on the number of different enzyme activities present, it may be desirable to partially purify the broth by selective precipitation of the preferred enzyme activity. The precipitated enzyme is then dissolved in an appropriate liquid. The concentrated solutions may be spray-dried to produce a concentrated powdered form of the enzyme or left as a liquid. To stabilize the enzyme, liquid enzyme is diluted to a standardized enzyme activity (usually based on the main enzyme activity present) with a liquid such as glycerol or corn syrup. Concentrated powders are mixed with diluents such as starch or lactose, again to a standardized activity level. These diluents act as carriers for the enzyme. The enzyme is concentrated so that very little (in terms of weight and volume) is required for use. Mixing the concentrated enzyme with a much larger volume of carrier makes measurement of the required amount of enzyme more convenient.

Enzyme producers are always looking for new sources and types of enzymes. They have looked all over the world by examining soils and other sources for different microorganisms and the enzymes they can produce. This is a time-consuming process and one that must be done very carefully to avoid unintended cross-contamination. Improved strains may also develop as the result of spontaneous genetic mutation of the organism's deoxyribonucleic acid (DNA), as well as from transfers of DNA from one microorganism to another. In any case, the goal is to constantly improve the microbiological sources for industrial enzymes.

When a promising new organism is discovered, a series of experiments must be conducted to determine how to refine and increase the organism's ability to produce the desired enzymes under indus-

Ultrafiltration—A process that uses a semipermeable membrane to separate fractions based on molecular size.

Lysis—Breakdown of the cell wall of a microorganism, which releases its contents.

trial conditions. A number of considerations are taken into account when selecting new organisms: 1) the amount of desired product produced relative to other products, 2) the complexity of the growing conditions, 3) whether some undesirable characteristics are present in the process, such as production of toxic by-products, pathogenicity, or poor culture stability, and 4) the ease of extracting the desired activity from the cell.

Since the 1970s, new techniques involving genetic engineering have made it possible to introduce the genetic code for virtually any protein into the DNA of a microorganism. In the past, if an enzyme that was thought to be of commercial value was discovered in a microorganism that was considered unsafe as a source of food-grade enzymes, the opportunity to produce the enzyme was normally lost. Today, however, the genetic material coding for that enzyme can be extracted and transferred to the DNA of an organism known to be safe for food-enzyme production. It is also possible to transfer DNA from an organism that is not well suited to industrial production conditions to one with better production capabilities. Such techniques have resulted in the commercial availability of a number of new and useful enzymes that can be produced with great efficiency.

SOME SAFETY AND REGULATORY ASPECTS

Production of a raw material for human consumption is, of course, subject to numerous safety regulations. To start with, the organisms used to produce enzymes for use in foods are restricted to a relatively few bacteria, yeasts, and fungi that have been tested over time and are considered to be safe. A partial list of these organisms, the primary enzymes produced from them, and their regulatory status is given in Appendix C. Currently, recognized enzymes produced from *A. niger, A. oryzae, B. subtilis*, and certain other microorganisms are generally recognized as safe (*GRAS*) and require no further approval from the U.S. Food and Drug Administration (FDA).

An exception to this is an enzyme activity that has not been previously recognized by the FDA. For example, phytase, the function of which is to degrade phytin to orthophosphorus, was not automatically recognized by the FDA and required approvals, even though it was derived from an *A. niger* organism. A new enzyme type, even though it is from an approved organism, requires safety studies and checks, at a minimum, before a GRAS petition can be filed. In addition, its fermentation process, including the growth media, cannot contain any components that might prove hazardous to human health. Safety checks are required on all organisms to ensure that their taxonomic characteristics are stable and have not been altered over time. This includes strains that may have foreign DNA inserted into their genetic material.

Enzymes derived from organisms not already approved for food use require approval, or at least a letter of nonobjection, from the FDA. Self-affirmation is now recognized by the FDA as an alternate

GRAS—A Food and Drug Administration regulatory status meaning Generally Recognized As Safe.

Good Manufacturing Practices—Defined practices for ensuring safety and accountability in manufacturing processes.

vehicle for approval. However, approval for food use may be granted only after documentation of extensive testing to demonstrate the safety of the new preparation. This process (Box 2-1) can be expensive and very time-consuming.

All aspects of the manufacturing process are recorded and checked on a regular basis to ensure compliance with *Good Manufacturing Practices* (GMP). All additives to the growth media must be food grade. The levels of extraneous microorganisms in the ingredients making up the media formulation are monitored, as is the level of heavy metals. Enzymes, in general, are added to food preparations in very small quantities, so the likelihood of problems arising from the enzyme itself is small. Nonetheless, GMP requirements must be strictly followed.

In general, enzymes do not function in the finished food product. They normally serve only to alter the processing conditions and/or the characteristics of the final product. For that reason, they are usually considered processing aids, as opposed to active food additives, and so are not required to be listed on product labels.

Storage of Enzymes

Although the optimal conditions for storage may depend on the particular enzyme involved and its form (liquid or powder), in general, enzymes should always be stored in a cool, dry environment. This includes the environment of the transit from manufacturer to customer as well as any transport done by an end-user within and between production facilities. To have a reasonable chance of producing the best product, year in and year out, companies must have consistent control over their raw ingredients, particularly functional ingredients such as enzymes. Storage can be a real problem in small companies, where storage facilities may be rather sparse. Even in large companies, multiple production sites with different facilities for enzyme storage and different environmental conditions throughout the year may make it difficult to maintain enzyme activity.

TEMPERATURE AND MOISTURE LEVELS

Heat can denature an enzyme under a variety of conditions. The enzyme is probably most susceptible to the harmful effects of heat when it is in a water solution. If the liquid diluent has a very low water activity, as is the case for high-fructose corn syrup or glycerol, the enzyme is less susceptible to the harmful effects of heat. No matter how the enzyme is stored, repeated exposure to heat has the potential to adversely affect the enzyme preparation and, in turn, the end product.

Some production facilities have large refrigerated rooms used for storing a variety of raw ingredients and suitable for storing enzymes. A refrigerator (or several refrigerators, if necessary) specifically designated for enzyme storage is another possibility.

Box 2-1. The Role of the Federal Government in the Marketing of Enzymes for Food Use[1]

An issue that always faces a company that intends to market an ingredient for use in a food is whether that food ingredient is subject to review by either the U.S. Food and Drug Administration (FDA), the U.S. Department of Agriculture (USDA), or the U.S. Bureau of Alcohol, Tobacco and Firearms (BATF) before the ingredient can be used in food. As with all government regulations, the answer is seldom quick and direct, but, on balance, most food ingredients are subject to relatively little direct government regulation. To begin with, the three regulatory agencies divide up their control over food ingredients based on the jurisdictions they have been granted by Congress. BATF is concerned with the use of an ingredient such as an enzyme in the use of alcoholic beverages; USDA's jurisdiction goes to the preparation of meat and poultry products; and all other food enzyme uses are subject to FDA's authority under its adulteration and misbranding provisions, to ensure the safety of all food ingredients.

The FDA is the most important agency to the industry because BATF and USDA typically permit the use of an enzyme in the foods they regulate if FDA has officially recognized its use by issuing a regulation or advisory opinion. Until 1958, the Federal Food, Drug, and Cosmetic Act had not placed any controls over ingredients that could be used in food except those that adulterated the food or diminished its quality. Then Congress passed the Food Additive Amendments of 1958, which provided that if the substance being used in a food was not generally recognized as safe (GRAS), either by use in food prior to 1958 or by scientific procedures (i.e., published studies demonstrating the safety of the ingredient), the firm wishing to use the ingredient in food must demonstrate and prove its safety through the submission of a food additive petition. When substances used in food are the subject of minor modifications and changes, the issue becomes whether that ingredient requires the submission of a food additive petition and subsequent issuance of a regulation or whether it is GRAS.

The 1958 statute left the determination of the GRAS status of an ingredient up to the firm marketing the food ingredient. There was no obligation to notify the FDA before entering the marketplace if you believed your ingredient was GRAS. Therefore, FDA had no legal obligation to decide which ingredients were GRAS. However, by the late 1960s it became apparent that the industry and FDA would benefit from a clear understanding of what was GRAS. Therefore, FDA proposed by regulation a GRAS affirmation petition process that would allow a company to submit a GRAS affirmation petition if it wished to receive an official FDA determination. However, such a petition was not a premarketing requirement. The process had a number of safeguards, required public notice, and resulted in a final regulation in the *Code of Federal Regulations*. This turned out to be a labor-intensive process. As more GRAS affirmation petitions were filed, the agency fell further and further behind in its review and affirmations.

There have been two major issues for FDA concerning the introduction of ingredients into the marketplace. One was the ability of the FDA to know what was being placed in the marketplace for the first time. The second was FDA's ability to notify a particular manufacturer that it had failed to completely resolve all the issues in regard to an ingredient in making the determination to go to market. The industry's major issue was the ability to tell suppliers that FDA was aware that the ingredient was in the marketplace. Recognition of these issues resulted in the GRAS notification proposal published by FDA in April 1997 (62 Fed. Reg. 18938, 18960 [April 17, 1997]). Once finalized, the GRAS notification will replace the GRAS affirmation petition process. As currently proposed, a company must submit a summary of the data on which it is relying to be able to market the ingredient (i.e., past history of use before 1958 or scientific data to support general recognition). There will be no *Federal Register* notices and no *Code of Federal Regulations* citation. The end result of this process will be a

continued

letter from FDA acknowledging that it has received the submission, the filing of the submission in the public docket, and a listing of the ingredients that are the subject of notification either in the public docket or the FDA web page. While this concept is currently a proposal, FDA has stated that it will begin using the notification process immediately.

Therefore, an enzyme manufacturer who has a new enzyme or has derived its enzyme from a new source and who wishes to go into the marketplace must make a decision about whether the substance is GRAS. If the company decides that the substance is GRAS, it may enter the marketplace subject to the risk that FDA may disagree, which is not likely. If the manufacture decides that the purchaser is likely to request information concerning FDA's knowledge of the ingredient, the manufacture may wish to submit a GRAS notification. If the substance is not GRAS, then the enzyme manufacturer will have to file a food additive petition, which means that the ingredient cannot enter the marketplace until the agency completes its petition review process, which can take anywhere from two to six years.

Both the BATF and USDA have their own regulations concerning ingredients that can be used in the products they regulate. Historically, both of these agencies have required a final determination from FDA as to safety; they have looked for a final GRAS regulation or a food additive regulation showing that the FDA had determined that the ingredient was safe for use in food. In those circumstances in which FDA did not publish a final regulation, but merely published a notice as part of the GRAS process, the ingredient could not be used in meat, poultry, beer, or wine. Recently however, both agencies have been involved in discussions with food companies and are making independent determinations of whether or not the ingredient is appropriate for use in a particular category of food. What impact, if any, FDA's notification process will have on BATF and USDA's review of ingredients is not clear, particularly since the FDA proposal did not acknowledge the reliance that BATF and USDA have placed in FDA's determinations in the past. Therefore, the enzyme manufacturer who makes a self-determination as to the use of an enzyme in food will likely have a great deal of difficulty in seeing that enzyme used in a meat or poultry facility or in a beer or wine. In the past, even if the manufacturer submitted a GRAS affirmation petition and it was accepted for filing, there still was a substantial hurdle to using the ingredient in meat and poultry or beer and wine. In fact, FDA's failure to finalize GRAS regulations had reached such a level that a number of manufacturers filed food additive petitions, not because they thought the substance was not GRAS, but because the main use was in either a meat and poultry or beer and wine product and they needed a final regulation to have the ingredient accepted for use in that particular category of food.

A different question concerns the use of the enzyme. Is it a processing aid, which means that the enzyme is not really an ingredient of the final food and is not required to be listed on the label, or is it an ingredient in the final product, which therefore must be included in the label ingredient list? Again, there are often subtle differences affecting when an enzyme is required to be listed in the label and when it is determined to be a processing aid and not part of the label.

In the final analysis, it is the responsibility of the manufacturer to ensure that the enzyme it proposes to market is safe. The manufacturer may make that determination on its own, or it may inform FDA of that determination. If the agency disagrees with the determination, a food additive petition will be required, which means that the ingredient cannot be marketed until the petition is approved. The question of whether the enzyme will appear on the final product label is usually not one for the enzyme manufacturer but for the end user. In our market system, determination of the enzyme's use has always been the responsibility of the manufacturer of the finished food product, and the food regulatory system has not changed that.

[1] This section was contributed by Gary Yingling of McKenna & Cuneo, L.L.P., counsel for the Enzyme Technical Association, Washington, DC.

Depending on the physical form of the enzyme, i.e., liquid or powder, several practices can help to preserve the original activity level. First, the enzyme should be packaged in moisture-proof containers. For powdered enzyme, the container should have an effective moisture barrier built into the packaging to prevent ambient moisture (humidity) from migrating through the package and into the enzyme. Most of the material in a package of powdered enzyme is a carrier such as starch or sugar (Table 2-1). The carriers may be hygroscopic, absorbing water from the atmosphere. If the packaging does not contain an effective moisture barrier, the water content of the enzyme/carrier mix can increase, making the enzyme more susceptible to the detrimental effects of heat.

Second, it is better to maintain a constant storage temperature (as close to optimal as practical) rather than to allow the enzyme (whether liquid or powder) to undergo repeated changes in temperature. Large fluctuations in temperature, especially if accompanied by corresponding changes in humidity, can be very harmful to the enzyme activity. In addition, the difference between the temperatures of the long-term storage area and the area where the enzyme is actually used should be minimized. If this cannot be accomplished, it is best to remove from storage only that amount of enzyme that will be used within a relatively short period of time. As a rule of thumb, the amount of enzyme removed from storage for use in production should never exceed the amount used during one production shift (8 hr).

Third, to expose the least amount of enzyme to nonstorage conditions, the enzyme should be packaged in relatively small amounts as opposed to large bulk containers. A corollary to this rule is that the number of handling operations involving enzymes in the production of a product should be minimized.

TABLE 2-1. Major Ingredients in Commercial Enzyme Production[a]

Constituent	Percent (w/w)	
	Powder	Liquid
Enzyme (active protein)	1–8	0.5–5
Carriers	Up to 90+	...
Preservatives	...	0.2–1.0
Electrolytes (e.g., NaCl)	2–5	Up to 16–18
Water	3–5	Up to 90+

[a] Adapted from (1).

> **Enzyme Storage Rules**
> 1. Use moisture-proof containers.
> 2. Keep refrigeration temperature constant.
> 3. Package in batch-sized units.

MAINTAINING ENZYME ACTIVITY

All enzyme users should be concerned about the length of time that an enzyme may be stored and still retain its full activity. Unfortunately, a precise answer is not always readily available. The answer depends on a number of factors, including the form of the enzyme (liquid or powder), the normal long-term storage conditions (again, the cooler and drier, the better), the susceptibility of the enzyme itself to environmental conditions, and the actual makeup of the enzyme preparation (carrier plus enzyme). The question is best answered by the end-user by developing and maintaining a quality control program at the production facility and monitoring the enzyme activity over time to determine when, under the specific storage conditions, the enzyme loses sufficient activity to adversely affect the production process. The tests discussed in Chapter 4 can be part of such a quality control system. In general, it is best to maintain the least amount of enzyme possible in inventory for the shortest period of time.

In spite of these precautions, it is not always possible to order the enzyme such that the current supply is always used well within the time determined for retention of full activity. The user may be faced with a decision about whether to use enzyme that has either gone past its expiration date or displays a significant loss of activity. The fact that the enzyme has passed its recommended use date should be of concern but may not be cause for immediate discarding of the enzyme. If no detrimental effects are observed in the product, the enzyme may continue to be used for a short time with cautious monitoring of the production process. New enzyme should be ordered as soon as possible. In general, if there is no independent means for monitoring enzyme activity (other than the assay provided by the manufacturer when the enzyme was shipped), it is best to discard the out-of-date enzyme and assume it is no longer usable. Although some usable enzyme may be discarded, the cost of producing out-of-specification product, plus any possible down-time, must be weighed against the cost of the discarded enzyme.

Handling and Use of Enzymes

In discussing appropriate methods for handling and using enzymes, it is useful to review some of the information already presented. Keep in mind that enzymes require a certain environment in which they can operate effectively. This environment is defined primarily by pH and temperature. Extremes of either one can result in significant loss of activity. It is useful to consider that enzymes are "comfortable" under conditions in which humans are most comfortable too. In other words, if the water in which you are about to dissolve the enzyme is too hot to be comfortable to you, it won't be comfortable for the enzyme either. Water that is too hot will destabilize the enzyme and lead to loss of activity. The same is true for extremes of pH. High pH (>8.0) or low pH (<5.5) may lead to enzyme denaturation. Be sure to check the literature accompanying the enzyme preparation to be certain of the conditions that constitute an appropriate "comfort range." Physical abuse, such as vigorous mixing or shaking, can also lead to loss of activity, as can the combination of any of these conditions.

The handling of enzymes varies tremendously within the food industry, but some guidelines will enable any company to devise an appropriate set of directions for handling and using enzymes for their specific environment. Handling of the enzyme involves all the steps required to get the enzyme from storage to the product makeup site and ready to be added to the product mix. The number of steps in this process should be minimized to avoid potential problems. The larger the quantity of enzyme in a unit package, the more difficult it will be to minimize the steps. The enzyme should be packaged such that one unit contains the appropriate amount of enzyme for one batch of product. This eliminates weighing the enzyme as well as ensuring that there will be no open, partially filled packets of enzyme that cannot easily be resealed and thus protected from changes in humidity and temperature. Also, only the number of packets required for production need be transported to the makeup site. If the enzyme liquid or powder is exposed to situations in which microbial contamination could occur, or where heat or other factors could cause the protein to be denatured, the enzyme will lose its activity.

If the enzyme is in liquid form, several alternatives can be considered. The most desirable is to have the correct amount of liquid enzyme automatically metered into the mixer. This eliminates the need for personnel to handle the enzyme at all during mixing. If this is not practical at a facility, the supplier should be asked to package the enzyme in a liquid carrier compatible with the formulation and in disposable containers holding an amount suitable for use in one batch. The additional packaging required for individual batch use may increase the cost of the enzyme as a raw ingredient, but this cost must be balanced against the convenience and accuracy of use. If this is not practical either, the user should make sure that the container is easily transported and that the contents can be conveniently poured out for weighing or volumetric measurement. A separate area should be designated for such weighing or measurement. The area should be cleaned thoroughly after each measurement of enzyme to prevent contamination between samples and minimize the chances of transfer of enzyme to the operators.

Once the amount of enzyme to be used has been correctly prepared, its use in the formulation depends on the form of the enzyme and the specific procedure called for. For example, when using a liquid enzyme, one usually can add the enzyme directly to the mix. Sometimes it is diluted with additional water to facilitate the homogeneous distribution of the enzyme throughout the entire batch. If the enzyme is in powder form, addition is a bit more involved. Unless the mix to which the enzyme is to be added is already a liquid, the powdered enzyme must usually be dissolved in a liquid, normally water, to ensure complete mixing within the product formulation.

SOME PRACTICAL EXAMPLES

When an enzyme is placed in a situation in which it can catalyze a chemical reaction, it will immediately proceed to do so. The following examples show improper handling.

Example 1. The person doing the mixing is a little behind schedule. She must prepare an enzyme solution from powdered enzyme and add it to an upright mixer. She fills a bucket with hot water, not checking the temperature of the water. She then adds the powdered enzyme all at once and, using a mixing device, stirs the solution vigorously. She notices that a foam has appeared on the surface but continues to mix vigorously. She then waits for the mix cycle to stop so she can open the mixer and add the water-enzyme solution to the formulation.

In this example, the operator has made two classic mistakes. The combination of water that is too hot and mixing that is too vigorous will very likely lead to a decrease in the enzyme activity. Thus, the amount of active enzyme actually added to the mixer may well be significantly less than that required to produce an optimal product. To correct this, always be sure that the water temperature used in making up the enzyme solution is not too hot (it will rarely be too cold). You can determine this by using your hand to test the water in which you plan to dissolve the enzyme. If it feels comfortable to you (this will generally be in the range of 60–85°F or 15–30°C), then it will be comfortable for the enzyme.

To efficiently mix the enzyme into the water, add it a little at a time while stirring gently. The enzyme will dissolve readily, without foaming.

Example 2. Some products require the addition of several solutions to the mixer. In this case, one of the solutions is the enzyme; the other is ammonium bicarbonate. The operator first prepares the enzyme solution in one bucket, then, in a second container, places the required amount of ammonium bicarbonate followed by water. He mixes the ammonium bicarbonate around and adds it to the mixer. He observes that only part of the ammonium bicarbonate has dissolved and that some of the solid ammonium bicarbonate has remained in the bottom of the bucket. To rinse the remaining ammonium bicarbonate out of the bucket, he pours the enzyme solution into the remaining ammonium bicarbonate, swishes it around, and adds the liquid to the mixer.

In this example, the enzyme may be denatured before it even gets into the mixer. The pH of the ammonium bicarbonate is quite basic (~8.2), and extremes of pH tend to destabilize enzymes and lead to denaturation. Thus, ammonium bicarbonate, a commonly used ingredient, can result in loss of enzyme activity if not handled properly.

Example 3. A formulation calls for adding two separate enzymes to the same product batch. The operator decides that it would be more efficient to add both enzymes to the same bucket of water.

However, this practice may have very detrimental effects on the activity of the enzymes involved. In general, it is not a good idea to mix enzymes together in the same solution, especially if one or both is a protease. While some proteases are very particular about what proteins they hydrolyze, most commercially available proteases are not at all choosy about what proteins to attack. If mixed with any other enzyme, the protease can attack and inactivate that enzyme. In fact, one protease enzyme in solution can attack and inactivate itself. Enzymes, especially proteases, should not be mixed with other enzymes in water solution and, when dissolved in water, should be added to the formulation as soon as possible.

Example 4. Having added powdered enzyme to a bucket of water, the operator gives the enzyme a couple of swishes with a whisk. There are still some pretty big clumps of enzyme floating in the water, but it is time to add the enzyme. The operator dumps the bucket into the mixer.

This example is also a commonly observed situation. The operator has not properly dissolved the enzyme, and the problem will not be as much with the enzyme itself as with the product. The amount of enzyme, either liquid or powder, added to a formulation is a very small percentage. For example, it may vary from about 0.005 to about 0.025% of the flour weight in a formulation. It is difficult to mix this amount of enzyme uniformly throughout the entire mix. However, a liquid is much easier to mix uniformly. That's why it is so important to add the enzyme in the form of a liquid. Leaving the enzyme in clumps, as this operator did, results in uneven distribution. This will lead to a poorly conditioned dough, which is likely to result in machining problems as well as poor product consistency. The enzyme must be properly mixed into a uniform solution and then the liquid enzyme preparation must be added to the mixer as soon as possible.

SOME SAFETY PRECAUTIONS

Humans are composed of a lot of protein, including exposed portions such as skin. Enzymes, particularly proteolytic enzymes, can attack and irritate human skin, especially the mucous membranes of the nose and eyes. The more concentrated the enzyme solution is, the greater is the potential for skin irritation. While nonproteolytic enzymes (e.g., amylases and pectinases) are much less likely to cause problems, it is best to minimize any contact of enzymes with the skin, eyes, and nose. Most commercial enzymes are not pure preparations containing only one type of activity. A commercial nonproteolytic enzyme preparation could, and often does, contain proteolytic activity.

Enzymes can cause irritation when they have prolonged contact with the skin or when powdered enzymes or enzyme-containing aerosols are inhaled. Liquid enzymes often include surfactants or penetrants that allow easier passage of the enzyme through the mucus membranes. Enzyme powders may be very fine and can spread easily through the air. Irritation is caused by the chemical properties of the enzyme and is not an allergic reaction.

However, some people can also suffer severe allergic reactions to enzymes. This can be a particular problem for people who already suffer from allergic responses to other allergens (e.g., have hay fever). Current evidence indicates that allergic symptoms result from breathing aerosols and not from skin contact. Symptoms of an allergic response may include any of the following: itching watery eyes, sneezing, runny nose, coughing, congestion, and difficulty in breathing. If any of these symptoms are experienced, a physician should be consulted as soon as possible.

Handling practices. It is extremely important when working with enzymes not to allow direct contact with the skin. Operators must take precautions so as not to spill the material or to breathe in any of the dust that may arise from powdered proteins. This requires the use of careful work practices and a proper ventilation system. High-energy operations that can generate aerosols, such as mixing, grinding, and using high water pressure or compressed air, should be avoided or done in contained areas.

A possible option when powdered enzymes are required is to use "dustless" preparations. These enzyme powders have been treated in such a way that they are much less likely to produce airborne particles that may enter the eyes or respiratory tract.

Spills should be cleaned up immediately with soap and water. This includes both the surface on which the enzyme was spilled and any part of the operator that may have come in contact with the enzyme. Plenty of water should be used to flush away the enzyme material. To minimize airborne particles, enzymes should not be removed by sweeping or with high-pressure water, air, or steam.

Protective equipment. Protective clothing, including rubber gloves, eye protection, and outer garments such as a lab coat, should be worn when there is a potential for enzymes to come in contact with skin or eyes. They are particularly necessary during operations such as spill cleanup or cleaning and repair of equipment. Protective clothing should be removed before the operator leaves the work area and should not be worn home. Respiratory protection, such as a breathing filter mask, is necessary for operations that might generate aerosols.

While enzymes, in general, are safe ingredients to work with, safety is a function of the user's understanding of the materials and procedures for handling them.

Reference

1. Godfrey, T., and Reichelt, J. 1983. *Industrial Enzymology: The Application of Enzymes in Industry.* The Nature Press, New York.

Supplemental Reading

1. Enzyme Technical Association. 1989. *Working Safely With Enzymes.* The Association, Washington, DC.
2. FAO. 1995. Codex Alimentarius. Secretariat of the Joint FAO/WHO Food Standards Programme, Food and Agriculture Organization, Rome.
3. Schwimmer, S. 1981. *Source Book of Food Enzymology.* AVI Publishing Co. Inc., Westport, CT.
4. Zapsalis, C., and Anderle Beck, R. 1985. *Food Chemistry and Nutritional Biochemistry.* John Wiley and Sons, New York.

CHAPTER 3

Common Enzyme Reactions

The topic of enzyme reactions is very broad; therefore, this chapter highlights some general reaction patterns common to food systems. The chapter is divided into sections on enzymes that hydrolyze carbohydrates, enzymes that hydrolyze proteins, enzymes that hydrolyze fats, and other enzymes or enzymatic reactions that are important in food systems.

Enzyme-Substrate Interactions

Basically, enzymes can hydrolyze a polymeric substrate in two ways. *Exoenzymes* can sequentially remove single polymer units from the end of polymer chains, whereas *endoenzymes* can rupture the internal bonds in a random manner at any point along the chains (Fig. 3-1).

Exoenzymes exhibit a significantly different effect on polymer viscosity than that of endoenzymes. A large polymer that is able to absorb and/or dissolve in water forms a viscous (thick) liquid. The viscosity is, in general, directly related to the size of the polymer. Because an exoenzyme can remove only single units of the polymer, one at a time, from one end, the viscosity of the polymer solution is changed very little by the action of exoenzymes. However, with a single clip in the middle of a large polymer, an endoenzyme can make two pieces that are only half the size of the original polymer. Therefore, the endoenzyme can reduce the viscosity of the polymer solution drastically. Figure 3-2 illustrates the effect of both forms of enzyme activity on the viscosity of a substrate solution.

Enzymes That Hydrolyze Carbohydrates

STARCH CARBOHYDRASES

Starch is made up from units of the sugar glucose (Fig. 3-3A). The numbers 1–6 represent the standard numbering system used to identify the carbon atoms in sugar molecules.

In starch, most of the glucose molecules are linked together by α-1-4 bonds, which connect the number 1 carbon atom of one glucose unit with the number 4 carbon atom of the adjoining glucose unit (Fig. 3-3B).

In This Chapter:

Enzyme-Substrate Interactions

Enzymes That Hydrolyze Carbohydrates
 Starch carbohydrases
 Cellulases
 Other carbohydrases

Enzymes That Hydrolyze Proteins
 Specificity
 Classification
 Action on protein

Enzymes Affecting Fats and Oils
 Lipases and other esterases
 Lipoxygenase

Other Enzymatic Reactions
 Oxidation and rancidity
 Enzymatic browning
 Enzymes involved in fruit ripening
 Other reactions

Exoenzyme—An enzyme capable of splitting only terminal bonds in a polymer chain.

Endoenzyme—An enzyme capable of splitting bonds anywhere along a polymer chain.

There are two kinds of starch polymers. *Amylose*, a linear molecule, is composed entirely of α,1-4 bonds. *Amylopectin*, a branched molecule, has the same chemical composition as amylose but has α,1-4 and α,1-6 bonds, which result in branching in the amylopectin polymer (Fig. 3-3C).

Amylases are enzymes that hydrolyze both amylose and amylopectin, but the extent of hydrolysis of the amylopectin is different because of the α-1,6 branching. The endoenzyme, α-amylase (EC 3.2.1.1), which is present in both plants and animals and is generally derived commercially from *Bacillus subtilis* or a related organism, is capable of attacking any internal α-1,4 bond in the starch chain. When amylose is hydrolyzed, the hydrolysis continues until the chain lengths are about 10–20 glucose units. After this, the starch fragments do not bind well to the enzyme and the α-amylase cannot further degrade the fragments. The result of α-amylase hydrolysis of amylopectin is a mixture of linear fragments, as for amylose, but also larger fragments that contain the α,1-6 bond, which cannot be broken by α-amylase.

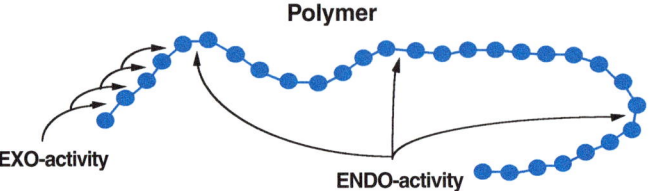

Fig. 3-1. Modes of endo- and exoenzyme attack on polymers.

Amylose—The type of starch molecule that occurs as a linear coil with no branching.

Amylopectin—The type of starch molecule that has branches.

β-Limit dextrin—The oligosaccharides resulting from the limited action of β-amylase (i,e., it cannot hydrolyze the 1,6-α-D-glucosidic linkage) in amylopectin.

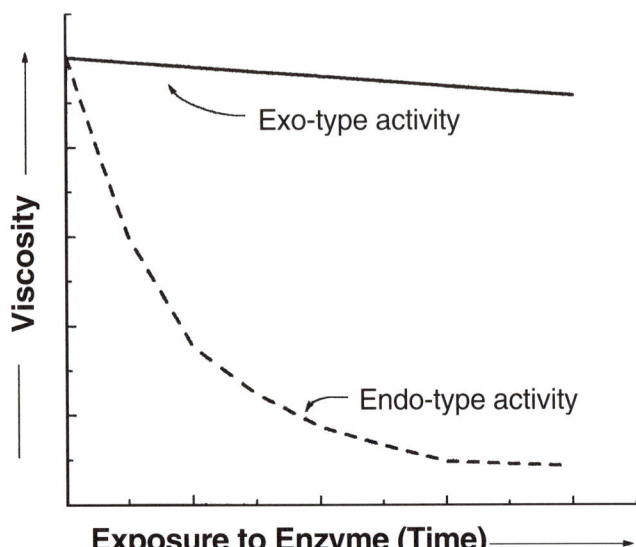

Fig. 3-2. Effect on viscosity of a polymer solution resulting from exo- and endo-enzyme activity.

The exoenzyme, β-amylase (EC 3.2.1.2), hydrolyzes starch by breaking the alternate (second) α,1-4 bond encountered as it moves along the chain from the nonreducing end, producing maltose. The reducing end of the starch polymer is the end that has a free hydroxyl group (OH) available on the no. 1 carbon atom of the terminal glucose group. Like α-amylase, β-amylase is unable to move through an α,1-6 bond and is therefore blocked from continuing when it encounters one. The result of hydrolysis of amylopectin is maltose plus the rest of the amylopectin molecule from the α,1-6 bond to the reducing end of the polymer. This larger starch fragment is generally referred to as a *β-limit dextrin*.

Several other enzymes, sometimes called *debranching enzymes*, can further hydrolyze starch and are very important for many industrial applications. For example, to efficiently produce high-fructose corn syrup from starch, the starch must be completely reduced to glucose so that another enzyme (a glucose isomerase) can convert that glucose to fructose. The hydrolysis of starch (both amylose and amylopectin) is enhanced by the action of debranching enzymes. The enzyme *pullulanase* (EC 3.2.1.41) is capable of hydrolyzing the α,1-6 bond found in amylopectin. *Amyloglucosidase*, or *glucoamylase*, (EC 3.2.1.3) converts the maltose and larger fragments resulting from the action of α- and β-amylase to glucose. For a diagrammatic review of the action of all the starch carbohydrases, see Figure 3-4.

CELLULASES

Cellulose is also a polymer of glucose. However, the bonds between the glucose units are β,1-4 bonds (Fig. 3-5).

The seemingly small alteration in the orientation of the bonds in cellulose as compared to those in starch make a big difference in the physical structures of the two glucose polymers and in their susceptibility to enzymes. The α-1,4 linkage imparts a coiled, helical structure to the amylose starch, whereas the β-1,4 linkage results in an extended linear cellulose polymer. Amylases can hydrolyze starch but cannot hydrolyze cellulose at all. *Cellulases* are required to break down cellulose but have no effect on starch. This illustrates the selectivity of enzymes in terms of their ability to bind to and hydrolyze only specific substrate molecules with a unique spatial orientation.

Cellulase enzymes can be either endo- (EC 3.2.1.4) or exo-cellulases (EC 3.2.1.5). Unlike starch, cellulose comes in several chemical forms and requires a significantly larger number of enzymes and processes to break it down to glucose. For example, cellulose is converted from the crystalline form to an amorphous state, followed by conversion to cellobiose. Finally, cellobiose is converted to two glucose units. In this processes, endoglucanases hydrolyze the amorphous regions

Debranching enzyme— Enzyme capable of hydrolyzing the branch points (e.g., the 1,6-α-D-glucosidic linkages) found in carbohydrate molecules such as amylopectin.

Pullulanase—A debranching enzyme capable of hydrolyzing the 1,6-α-D-glucosidic linkages in carbohydrates such as pullulan and amylopectin.

Amyloglucosidase (glucoamylase)—An enzyme that hydrolyzes both 1,4-α-D- and 1,6-α-D-glucosidic bonds in carbohydrates, although at different rates.

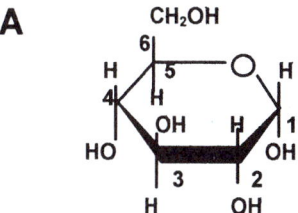

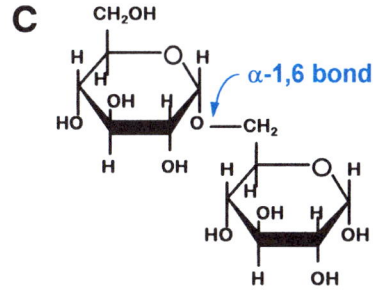

Fig. 3-3. Structure of glucose (A). Formation of maltose from two glucose molecules (B). The α-1,6 bond linking two glucose molecules (C).

Cellulase—An enzyme that hydrolyzes cellulose, a polymer of β-1,4-linked glucose molecules, which is a common component of plant cell walls.

Xylanase—An endopentosanase enzyme that hydrolyzes the β-1,4 xylosidic bonds in the xylose polymer backbone of which pentosans are composed.

by random hydrolysis of β-glucosidic bonds. Cellobiohydrolases sequentially remove cellobiose from the nonreducing ends of the cellulose. Exoglucohydrases remove consecutive glucose units from the nonreducing end of cellodextrins. Then β-glucosidases cleave cellobiose to glucose.

OTHER CARBOHYDRASES

Other carbohydrate polymers important in the food industry include pentosans, β-glucans, and pectic substances. These carbohydrates are composed of different sugars than are found in starch or cellulose.

These substances, often falling into the nebulous category of hemicelluloses, are all highly branched, and the percentages of each vary substantially from plant source to plant source. The specific enzyme action by which they are hydrolyzed is often difficult to characterize on a commercial basis. Also, the enzyme preparations are often composed of many different side activities. However, the basic reaction of viscosity reduction or sugar production remains consistent.

Pentosanase. Pentosan, a natural constituent of cereal-based flours, consists of several sugars, primarily xylose and arabinose. Its structure typically consists of a backbone chain of xylose sugars linked by β,1-4 bonds. At certain points along this xylose backbone, five-membered arabinose sugars are found linked to the xylose sugar through α-1,3 bonds (Fig. 3-6). The brackets in this figure indicate that the two xylose units are repeated n number of times, where n can be any whole number.

The pentosan polymer absorbs about 10 times its weight in water and thus dramatically affects the consistency of dough systems. Hydrolysis of the pentosan network with an endopentosanase enzyme (*xylanase*, EC 3.2.1.32) greatly diminishes the water-holding capacity of the pentosan polymer. Pentosanases are also found in many plants.

β-Glucanase. The β-glucan polymer is typically found in oats, barley, and some fruits and is produced by some microorganisms. β-Glucan polymers are units of glucose linked by both β,1-3 and β,1-4 bonds (Fig. 3-7). These polymers contribute to the beneficial

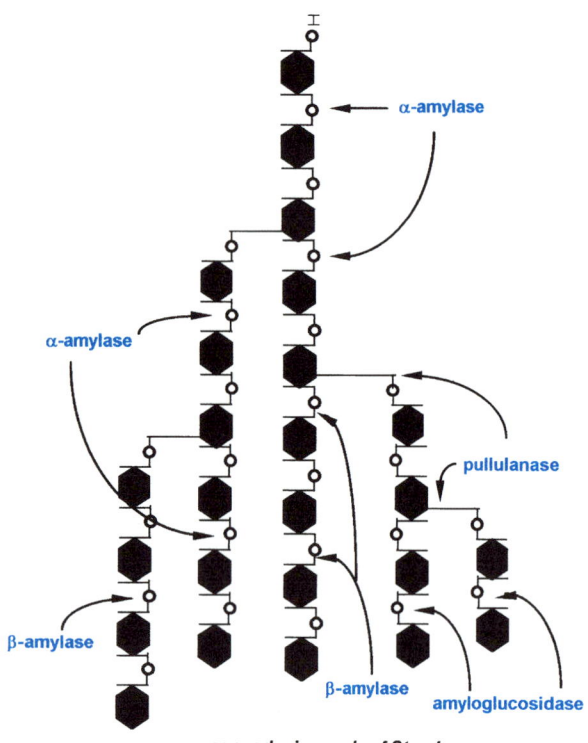

Fig. 3-4. Action patterns of starch-degrading enzymes.

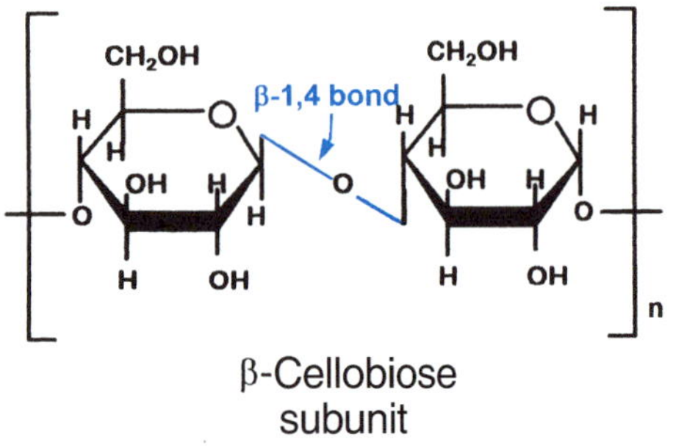

Fig. 3-5. The β-1,4 bond in the cellobiose subunit of cellulose.

soluble fiber in oatmeal and barley and also cause the problem of

instance, in the production of fruit juice, β-glucans in the fruit can increase the viscosity, making it difficult to filter the juice. β-*Glucanase* (EC 3.2.1.6), which degrades β-glucan, has been found to be very effective in eliminating problems associated with the high viscosity of the β-glucan polymer. In brewing, β-glucanase from malt or an exogenous source such as microbial enzymes is used to reduce viscosity. However, in some applications, high-viscosity β-glucans are considered beneficial, e.g., β-glucans in oatmeal are associated with lowering of cholesterol. In this case, the food producer would want to avoid β-glucanase activity, which could be a side activity from an enzyme preparation.

Pectinases. The term pectic substance refers to a group of colloidal carbohydrates in plants that are constituents of plant cell walls and intercellular layers and contribute to the textural qualities of many fruits and vegetables. Pectic substances are polymers composed of three monosaccharides: D-galacturonic acid, D-galactose, and L-arabinose.

Pectin is composed of the methyl ester of α-1,4-linked galacturonopyranose units. Galactans are composed of β-1,4 linked D-galactopyranose units, while araban consists of α-1,5 linked L-arabinofuranose units with α-1,3 cross linkages. Examples of these sugar polymers are shown in Figure 3-8.

Protopectin is the genitive pectic substance in plants. Because some degradation of protopectin is required to extract it from the plant cells, its precise structure is not known.

Pectin is the pectic material that, when heated with sucrose, forms a clear gel widely used in the preparation of jellies and jams. High-methoxyl pectins have more than half of the carboxyl groups in the methyl ester form; those with less than half in that form are low-methoxyl pectins. Pectic acid is the polymer that results when virtu-

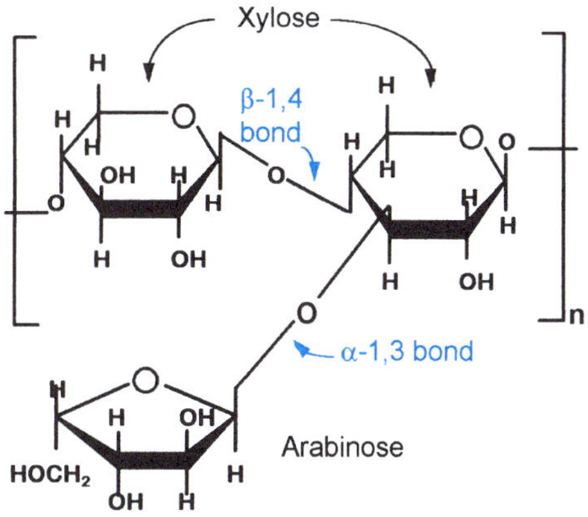

> **β-Glucanase**—An endoglucanase enzyme that hydrolyzes the 1,3- and 1,4-β-D-glucosidic bonds in β-glucan.

Fig. 3-6. Pentosan structural unit.

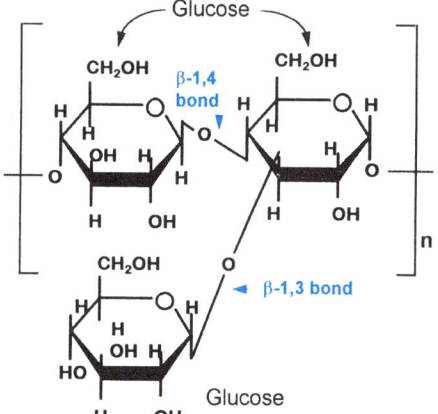

Fig. 3-7. β-glucan structural unit.

Lyase—An enzyme that removes specific groups from their substrates, usually by elimination rather than hydrolysis, leaving a double bond.

ally all the methoxy ester groups on pectin have been enzymatically hydrolyzed to carboxyl groups in the form of the free acid.

Three types of enzymes hydrolyze pectic substances. The first, pectin esterase (3.1.1.11), removes the methyl ester on the number 6 carbon of the methylated galactose units in pectin, resulting in pectic acid. This enzyme is also referred to as pectase, pectin demethoxylase, and pectin methylesterase.

Polygalacturonases split the glycosidic linkages between galacturonosyl units through a hydrolysis mechanism (Fig. 3-9A). Polygalacturonases have an endo form (EC 3.2.1.15) and two exo forms, one of which (EC 3.2.1.67) releases D-galaturonic acid. The other (EC 3.2.1.82) releases digalacturonic acid from the nonreducing end. Separate but similar forms of polygalacturonase act on either pectin or pectic acid. Those hydrolyzing pectin are called polymethylgalacturonases. While there is some evidence that methylated pectin is hydrolyzed at the 1,4 glycosidic linkage, an endoacting polymethylgalacturonase has not yet been isolated or characterized. The enzymes isolated thus far act only on pectic acid.

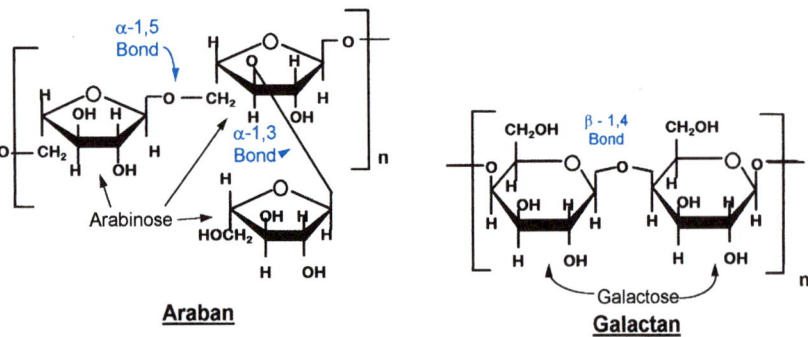

Fig. 3-8. Structural units of pectic substances.

The third category of pectic enzymes consists of *lyase* enzymes, which also split the galacturonosyl chain at the glycosidic linkage, but by *β-elimination* rather than by hydrolysis (Fig. 3-9B). The enzyme acting on pectin is called pectin lyase (EC 4.2.2.10). Only the endo form of this enzyme is currently identified. Lyases acting on pectic acid are called pectic acid lyases. Both the endo form, pectate lyase (EC 4.2.2.2), and the exo form, exopolygalacturonate lyase (EC 4.2.2.9), have been found. These enzymes are important in the processing of fruits and vegetables.

Enzymes That Hydrolyze Proteins

Enzymes that hydrolyze proteins are called proteases. The site of the attack is always at the peptide bond (also called the amide bond) between the amino end of one amino acid and the carboxyl end of the adjacent amino acid. As is the case with other enzymes that act on polymers, there are exo- and endoproteases. Endoproteases hydrolyze peptide bonds anywhere along the protein chain. The exoproteases, which attack the ends of the protein chain and remove one amino acid at a time, are called *carboxypeptidases* if they work from the carboxy terminus, or *aminopeptidases* if they work from the amino terminus.

SPECIFICITY

Proteases have an additional level of specificity. The specificity of the starch hydrolases depends on which kind of linkages between the individual glucose units they can hydrolyze. In proteins, the bonds between the amino acids are all peptide bonds, but the basic chemical units come from a pool of about 20 different amino acids, not just a single chemical unit like the glucose in starch. The various R-groups (see Appendix A) that are characteristic of amino acids add another level of complexity to protein polymers,

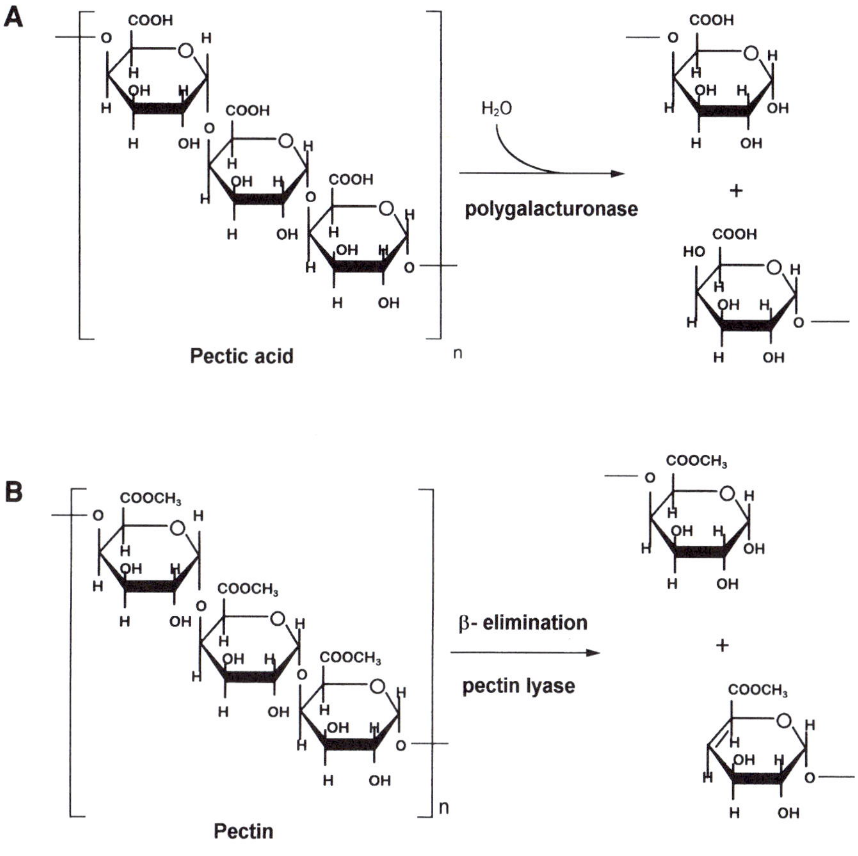

β-Elimination—A type of chemical reaction, not involving water, in which a specific group such as a hydrogen or hydroxyl is removed, resulting in the formation of a double bond.

Carboxypeptidase—Exoprotease hydrolyzing the amino acid on the carboxy end of a protein.

Aminopeptidase—Exoprotease hydrolyzing the amino acid at the amino end of the protein.

Fig. 3-9. Effect of two enzymes on pectic substances: polygalacturonase activity (hydrolysis) on pectic acid (A) and pectin lyase activity (β-elimination) on pectin (B).

requiring another level of specificity in protease enzymes. So, in addition to being either exo- or endo-, some enzymes also prefer to cleave peptide bonds involving specific amino acids. For example, trypsin (EC 3.4.21.4) is an animal endoprotease that hydrolyzes only peptide bonds in which an arginine or lysine amino acid residue is involved. This enzyme is said to have a very limited specificity in that it will hydrolyze proteins only at a limited number of sites, depending upon their amino acid contents. Other protease enzymes, such as papain (EC 3.4.22.2), are much less particular about the amino acids encountered. These enzymes are said to have broad specificity because they can hydrolyze most peptide bonds. Most of the protease enzymes available commercially are relatively nonspecific and can hydrolyze a large number of peptide bonds in the substrate proteins.

Table 3-1 illustrates this specificity. The substrate that was used for these data is a polypeptide chain from insulin. The table shows the number of peptide bonds broken in the substrate protein by different proteolytic enzymes after a certain period of time and identifies the specific amino acids making up the peptide bonds that were most readily hydrolyzed. The various protease enzymes show the ability to break different bonds in this substrate protein. Relatively nonspecific enzymes can hydrolyze a large number of peptide bonds, but they do not break them at the same rate. For this reason, the number of bonds broken shown in Table 3-1 does not necessarily correspond with the bonds shown as most readily split.

CLASSIFICATION

Most proteolytic enzymes can also be classified on the basis of the chemistry of their catalytic mechanism. Four groups are often designated: serine, thiol or sulfhydryl, metallo-, and acid. Figure 3-10 illustrates the operating pH ranges for some selected enzymes in these classes.

TABLE 3-1. Protease Specificity Against an Insulin B Chain[a]

Protease	Number of Bonds Aggressively Broken	Peptide Bonds Most Readily Split
Chymosin	2	Glu-Ala, Leu-Val
Trypsin	2	Arg-Gly, Lys-Ala
Pepsin	5	Leu-Val, Phe-Tyr
Bacterial		
Neutral	6	His-Leu, Ser-His, Ala-Leu, Gly-Phe, Arg-Gly
Alkaline	7	Gln-His, Ser-His, Leu-Tyr
Fungal		
Acidic	9	His-Leu, Gly-Phe, Phe-Phe
Alkaline	5	Leu-Tyr, Phe-Tyr
Papain	9	Asn-Gln, Glu-Ala, Leu-Val, Phe-Tyr

[a] Adapted from (1).

Serine. *Serine*, or *alkaline*, *proteases* require a hydroxyl function at the active site in order to function properly. Optimal activity tends to be in the alkaline pH range, usually between about 7.5 and 10.5. Trypsin is the classic example of this type of protease. Trypsinlike enzymes hydrolyze peptide bonds in which arginine or lysine provide the carbonyl group. Many proteases derived from bacterial sources, *Bacillus* species in particular, tend to be serine proteases.

Thiol. *Thiol proteases* require a sulfhydryl function at the active site, which is contributed by the amino acid cysteine. The pH range in which these enzymes function is usually rather broad, with greater activity below pH 7. An example of a thiol protease commonly used in food applications is papain, an enzyme recovered from the papaya plant.

Metallo-. *Metalloproteases* require the presence of a metal ion such as zinc at the active site. They tend to have pH optimal values around 7.0 and were originally called "neutral" proteases. Carboxypeptidases A and B (EC 3.4.17.1 and 3.4.17.2), as well as a number of proteases derived from fungal and bacterial sources, fall into this category.

Acid. *Acid proteases*, normally require carboxylic acid functions at the active site and are functional at a relatively low pH (2–3). The best examples are pepsin (EC 3.4.23.1), found in the human stomach, and *chymosin* (EC 3.4.23.4), from the digestive tract of young cows and goats. An example of a microbe-derived acid protease is from *Aspergillus niger*. This preparation has a pH optima of 2.5–3.5. Microbe-derived acid proteases can be derived from a variety of organisms, e.g., *A. meleus*.

Alkaline (serine) protease—A type of proteolytic enzyme, typically having a serine and a histidine residue in its active site and having activity in the alkaline pH range (about pH 7–11). Examples: trypsin, chymotrypsin, elastase.

Thiol protease—A type of proteolytic enzyme having a cysteine residue in the active site, operative over a wide pH range but typically between pH 4.5 and 9.5 with minimum around pH 6–7.5. Examples: papain, bromelain, ficin.

Metalloprotease—A type of proteolytic enzyme having a metal atom, usually Zn, at the active site and having a pH range that centers about pH 7.0. Examples: carboxypeptidase A and B, aminopeptidases, and dipeptidases.

ACTION ON PROTEIN

The viscosity of a protein solution depends on the size and shape of the protein polymer. The larger the protein, the more viscous the solution is likely to be. As with starch polymers, exoproteases have relatively little effect on the overall size of the protein and hence little effect on viscosity, while the action of an endoprotease can dramatically decrease the viscosity of the solution.

Although most industrial applications for proteases require endoproteolytic activity, to affect viscosity, exoproteases also have important roles. For example, exoproteases can enhance Maillard reactions by cleaving amino acids from proteins to make them available to participate with reducing sugars in browning reactions.

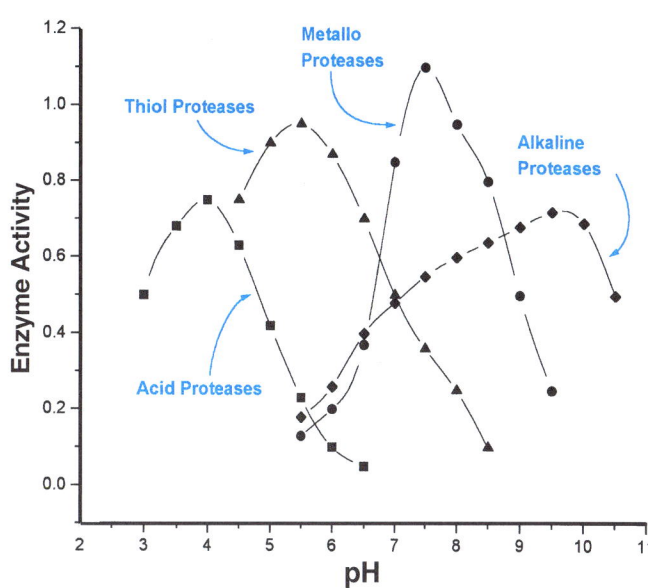

Fig. 3-10. pH ranges for classes of protease enzymes.

Acid protease—A type of proteolytic enzyme that has carboxylic acid functions (often from aspartic acid residues) at the active site and that functions in an acidic pH range (2–4). Examples: pepsin, chymosin, cathepsin D.

Chymosin—An acid protease, also called rennin, derived from the digestive tract of ruminants, used in the dairy industry.

TABLE 3-2. Effect of Proteases on Various Protein Substrates[a]

	Substrate Protein			
Protease	Hemoglobin	Gelatin	Casein	Soy Protein
Papain	100[b]	100	100	100
Trypsin	40	45	45	60
Bacterial				
Neutral	135	140	130	140
Alkaline	160	130	165	170
Fungal				
Acidic	65	20	80	25
Alkaline	20	65	95	70
Pepsin	65	70	95	10

[a] Adapted from (1).
[b] 100 is an arbitrary base number applied to the extent of hydrolysis by papain. All other numbers are relative to this base.

Simply bringing a protease into the presence of a protein does not necessarily lead to protein hydrolysis. The enzymatic reaction requires that *both* the enzyme and the substrate be in the proper state. For example, hemoglobin must be denatured before it can be a substrate for trypsin; otherwise, no proteolysis will occur. Hemoglobin, in its native state, is a very compact, spherical protein. The bonds susceptible to trypsin activity are buried in the interior of the protein's tertiary structure and are not accessible to the enzyme. After denaturation, the tertiary structure is altered such that the susceptible bonds become accessible.

In addition, because of specificity, an enzyme that is capable of hydrolyzing protein A may not be able to hydrolyze protein B. For example, a protease that can hydrolyze casein could be relatively ineffective against gluten protein. Table 3-2 shows the effectiveness of different proteases acting on different substrate proteins. The units shown are arbitrary, based on the activity of papain as a standard. In addition, a protease may be quite effective against casein (an animal protein) but not very efficient at hydrolyzing a plant protein such as soy. This specificity should be kept in mind when considering an appropriate enzyme for use in a particular process. Reaction rate is another factor to consider in the choice of enzymes. Slower enzymatic reactions may be better suited for specific applications or processing requirements.

Enzymes Affecting Fats and Oils

A fat or oil, also referred to as a lipid, exists as a compound called a triacylglycerol, which is a glycerol backbone to which three fatty acids are attached (Fig. 3-11). Fatty acids that are not part of a triacyl-

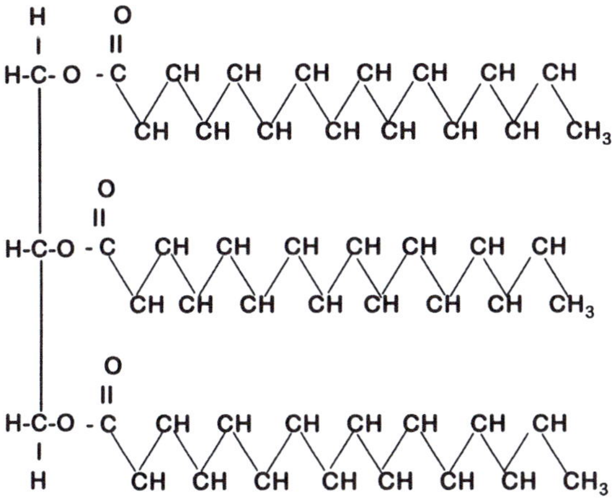

Fig. 3-11. Structure of triacylglycerol tripalmitin.

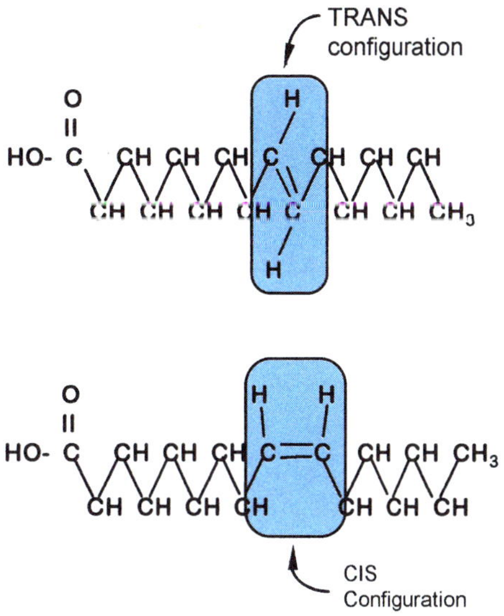

Fig. 3-12. The *trans* and *cis* structures in a monounsaturated fatty acid.

glycerol are referred to as free fatty acids or acylglycerols. The number of carbon atoms in the fatty acid chain can range from four to 24. Saturated fatty acids contain all single bonds between carbon atoms. Unsaturated fatty acids have one or more double bonds between the carbon atoms in the chain (Fig. 3-12). Monounsaturated fatty acids contain only one double bond, and polyunsaturated fatty acids have more than one. The double bond can be in a *cis* or a *trans* configuration. The *cis* form is more commonly found in nature, while the *trans* form, the less chemically reactive of the two, is more often the result of hydrogenation during commercial processing.

The enzymes of most importance in terms of their effects on fats and oils are *lipases* and other *esterases* and lipoxygenases.

LIPASE AND OTHER ESTERASES

Lipase (EC 3.1.1.3) attacks a triacylglycerol and removes the fatty acids from the glycerol backbone. Its pH optimum is generally 7–8. Most lipase enzymes remove the two outside fatty acids, leaving the middle fatty acid attached to the glycerol (Fig. 3-13). Thus, the result of enzymatic hydrolysis of a triacylglycerol by lipase is two free fatty acids and a monoglyceride. These free fatty acids tend to be chemically reactive, especially if they are unsaturated. Free unsaturated fatty acids can react with oxygen in the air, which leads to rancidity. Lipase is a normal component of many food substances and so is always of some concern in maintaining quality. It can also be used to affect the flavor characteristics of foods, since the free fatty acids released by lipase activity usually contribute more flavor than fatty acids in a triacylglycerol. This technology is used, for example, to

cis—Isomer in which both hydrogen atoms at a double bond are on the same side.

trans—Isomer in which the hydrogen atoms at a double bond are on opposite sides.

Lipase—An enzyme that hydrolyzes a triacylglycerol, forming free fatty acids.

Esterase—An enzyme capable of hydrolyzing an ester linkage, resulting in an acid and an alcohol.

create characteristic flavors in cheeses and enzyme-modified cheese flavors.

The action of the lipase esterase enzyme results in the production of mono- and diglycerides, which may be further hydrolyzed by the action of other esterases, specifically the carboxylesterases (carboxyl esterhydrolase, EC 3.1.1.1). These enzymes act on water-soluble esters and have been isolated from a variety of microorganisms. Under conditions in which the medium of reaction is a water-alcohol mixture, the carboxylesterases, as well as arylesterases (EC 3.1.1.2) from yeast, can synthesize ester compounds that may contribute to both desirable and detrimental flavor characteristics in dairy-based food products.

LIPOXYGENASE

Lipoxygenase (EC 1.13.11.12) is a metalloenzyme (containing an iron atom) that catalyzes the reaction of oxygen with a certain type of fatty acid, one that contains a structural unit called a *cis-cis* penta-1,4-diene unit (i.e., $-CH=CH-CH_2-CH=CH-$), sometimes referred to as a *cis-cis*-methylene-interrupted fatty acid. Thus, the preferred substrates for lipoxygenase are linoleic (9,12-octadecadienoic acid), linolenic (9,12,15-octadecatrienoic acid), and arachidonic acid (5,8,11,14-eicosatetraenoic), all commonly found in plants and animals. However, only the isomers of these fatty acids containing the *cis-cis*-penta-1,4-diene unit are able to serve as substrates for lipoxygenase. The enzyme catalyzes the addition of oxygen to the fatty acid at the site of unsaturation (the double bond), forming a reactive compound called a hydroperoxide. The result of this reaction is the oxidation of the unsaturated fat or oil.

There are two types of lipoxygenase. As illustrated in Figure 3-14, the type 1 lipoxygenase reaction, which occurs only with the free fatty acid form, results in either the 9- or the 13-position hydroperoxide, depending on the source of the lipoxygenase. The type 2 lipoxygenase can react with esterified fatty acids (i.e., a triacylglycerol)

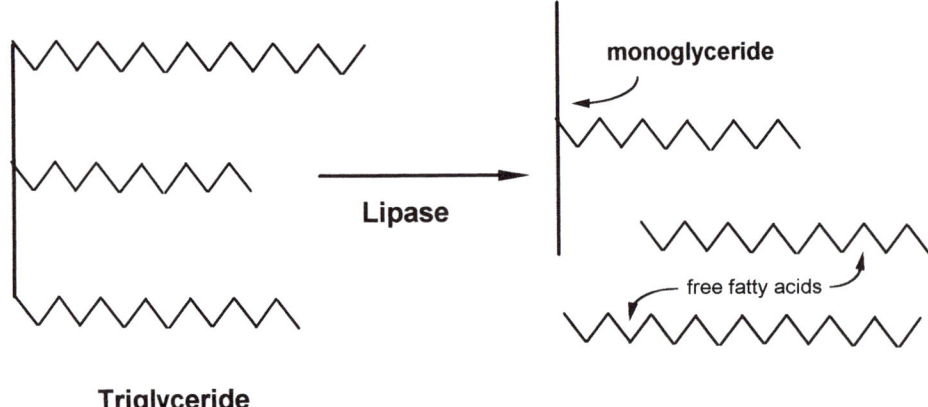

Fig. 3-13. Action of lipase enzyme on triacylglycerols.

and produces approximately equal amounts of the 9- and 13-hydroperoxides with linoleic acid as substrate. Type 2 lipoxygenase is also able to cooxidize carotenoids and chlorophylls, which results in a loss of color in these pigments. This activity is responsible for the "bleaching" effect of lipoxygenases from soy flour. Both forms of lipoxygenase are normal components of most living organisms, including most food raw materials.

Other Enzymatic Reactions

OXIDATION AND RANCIDITY

Lipid oxidation can be a major contributor to food spoilage or loss of quality. Oxidized or rancid off-flavors result from the addition of oxygen atoms across the double bond of unsaturated fatty acids. Factors that can accelerate or cause oxidation include light energy, temperature, moisture content, presence of metals, salt, and the structure of the fatty acids. Although oxidation reactions can proceed spontaneously in the presence of oxygen, lipoxygenases also can act to catalyze these reactions, forming free radicals that can further participate in the oxidation reactions. Prevention of rancidity is a concern in processing, packaging, distribution, and storage of food products.

Hydrolytic rancidity results from the hydrolysis of ester bonds in lipids, which may occur because of heat and moisture or because of enzymes, such as lipases or esterases. Heat treatment can be used to inactivate the enzymes; however, in some cheeses and chocolate, the liberation of free fatty acids is desired because they enhance the flavor profile.

ENZYMATIC BROWNING

The Maillard or nonenzymatic browning reaction, which turns baked goods brown, is a chemical reaction between an amino acid

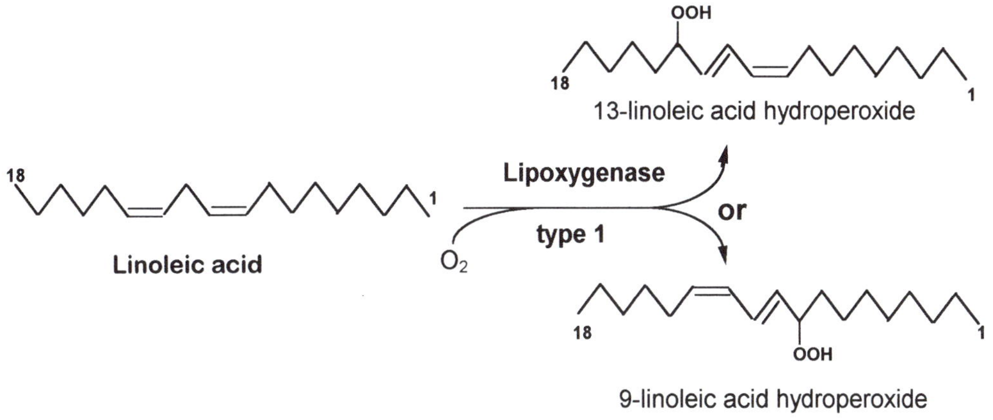

Fig. 3-14. Oxidation of the *cis-cis*-penta-1,4-diene unit in linoleic acid by type 1 lipoxygenase.

Enzymatic browning—Darkening of food products resulting from the action of polyphenol oxidases.

Polyphenol oxidases—A group of oxidative enzymes acting on phenolic substrates.

Superoxide dismutase—An enzyme that catalyzes a reaction between a superoxide anion (O_2^-) and hydrogen, producing hydrogen peroxide and molecular oxygen (O_2).

and a sugar and does not involve an enzyme. *Enzymatic browning* is an oxidation reaction involving a group of enzymes collectively called *polyphenol oxidases* (EC 1.10.3.1, 1.10.3.2, and 1.14.18.1), which are found naturally in most plant materials, including fruits, vegetables, and cereal grains. Under normal circumstances, the enzymes are held within the cells of the plant away from oxygen. However, when the fruit or vegetable is cut or bruised, the enzyme is released and reacts with certain phenolic compounds (such as catechol, substituted phenols, and the amino acid tyrosine) and with oxygen to produce brown-colored products. The reaction involving tyrosine (Fig. 3-15) produces melanin, a dark-colored pigment that can combine with proteins. These enzymes have also been implicated in the development of undesirable color formation in pasta products such as noodles. While some of this color is associated with wheat varietal differences, enzymatic activity is also believed to be involved in the color change.

Several methods are available for retarding or preventing color changes resulting from polyphenol oxidase enzymes. Antioxidants such as ascorbic acid can be used to inhibit this reaction, both through an antioxidant action and by lowering the pH. Since this enzyme is also a metalloenzyme requiring a copper ion, some sulfur compounds can be used to inhibit the enzymatic reaction. Sulfur binds very strongly to copper ions and makes them unavailable to the enzyme. Sulfur dioxide, sodium sulfite, and sodium hydrogen sulfite have been used, but this practice has been restricted because of reports of allergic responses to some sulfur compounds. Sulfite is now required to be listed on the labels of such products as wine and sulfite-treated fruits and vegetables. Heat treatment (to 95°C) may also be an effective means of inactivating these enzymes.

Another enzyme recently found to be involved in the enzymatic browning process in foods is called *superoxide dismutase* (SOD) (EC 1.15.1.1). This enzyme reacts with an activated form of oxygen, referred to as the superoxide anion (O_2^-), producing hydrogen peroxide and molecular (nonactivated) oxygen. SOD has been found to stabilize ascorbic acid, which inhibits the enzymatic browning reaction of the polyphenol oxi-

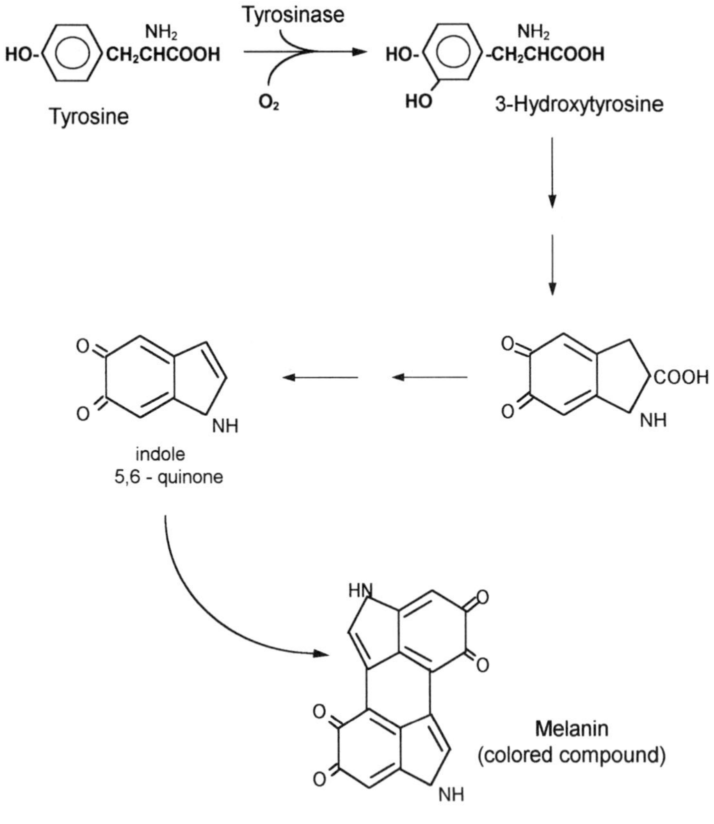

Fig. 3-15. Formation of colored compounds by the tyrosinase polyphenol oxidase enzyme.

dases. It can also significantly retard enzymatic browning in a number of fruits and vegetables, including mushrooms, apples, peaches, and potatoes.

ENZYMES INVOLVED IN FRUIT RIPENING

When fruits over-ripen, they usually tend to get softer, with accompanying textural changes such as "mealiness" and rubberiness. Most of these changes are the result of enzymes that affect the pectic material in fruits. Structurally, fruit cell walls consist of several carbohydrates, including cellulose and pectins. In green (unripe) fruit, the pectin is present as protopectin, an insoluble form that contributes to the hardness associated with unripe fruit. During the ripening process, pectin-degrading enzymes (pectinases) are released and attack the protopectin. This action results in degradation of the cell walls, which accounts for the softening. The enzyme activity of pectinase has been prevented in tomatoes through genetic engineering, resulting in delayed ripening and improved handling characteristics. Other enzymes, including amylases, proteases, and cellulases, are also activated during the ripening process. If present in excess, all can contribute to the undesirable softening and darkening of the fruit.

OTHER REACTIONS

Both *glucose oxidase* (EC 1.1.3.4) and *catalase* (EC 1.11.1.6) are used in some industrial applications. For instance, commercially dehydrated eggs are produced using glucose oxidase, which converts glucose to gluconic acid and hydrogen peroxide. This makes the glucose unavailable to participate with egg proteins in Maillard browning reactions. The reaction also helps to remove dissolved oxygen from the system. Catalase is added to convert the resulting hydrogen peroxide into water. In some cases, hydrogen peroxide is used as a sterilizing agent, and catalase can be used to break down excess hydrogen peroxide. Further heat treatment then inactivates the catalase enzyme.

Glucose oxidase—Enzyme that hydrolyzes glucose to gluconic acid and hydrogen peroxide.

Catalase—Enzyme that breaks down hydrogen peroxide to oxygen and water.

Reference

1. Godfrey, T., and Reichelt, J. 1983. *Industrial Enzymology: The Application of Enzymes in Industry.* The Nature Press, New York.

Supplemental Reading

1. Alais, C., and Linden, G. 1991. *Food Biochemistry.* Ellis Horwood, London.
2. Belitz, H. D., and Grosch, W. 1987. *Food Chemistry.* Springer-Verlag, New York.
3. Berk, Z. 1976. *Braverman's Introduction to the Biochemistry of Foods.* Elsevier Scientific Publishing Co., London.
4. Brockerhoff, H., and Jensen, R. G. 1974. *Lipolytic Enzymes.* Academic Press, New York.
5. Crooke, S. T., and Wong, A., Eds. 1991. *Lipoxygenases and Their Products.* Academic Press, New York.

6. Coultate, T. P. 1988. *Food: The Chemistry of Its Components*. Royal Society of Chemistry, London.
7. Fennema, O. W., Ed. 1985. *Food Chemistry*. Marcel Dekker, New York.
8. Schwimmer, S. 1981. *Source Book of Food Enzymology*. AVI Publishing Co. Inc., Westport, CT.
9. Whitaker, J. 1972. *Principles of Enzymology for the Food Sciences*. Marcel Dekker, New York.
10. Whitaker, J., and Tannenbaum, S. R. 1977. *Food Proteins*. AVI Publishing Co., Westport, CT.
11. Zapsalis, C., and Anderle Beck, R. 1985. *Food Chemistry and Nutritional Biochemistry*. John Wiley and Sons, New York.

Analysis of Enzyme Activity

Before discussing the various ways by which enzymes can be analyzed, it may be helpful to answer the basic question of why one would analyze enzymatic activity. Let's assume that you are involved in a quality assurance/quality control program at a food production facility. You are making every effort to produce the highest quality product as consistently as possible. This requires strict control over all raw ingredients, particularly the functional ingredients, the ones that contribute specific, desirable effects in the processing of your finished product. If control is not maintained, inconsistencies in both the processing and the finished product will result.

A number of questions might occur to you when thinking about how to achieve such control over enzymes. First, is the activity of the enzyme you are now using the same as the enzyme activity from the previous batch? Assuming that a *specification* has been set for the enzyme activity, does the activity of the enzyme meet that specification? How can you compare the activity level and functionality of enzymes from different suppliers? Are other enzymes present, in addition to that identified by the supplier, that can affect the production process? How long can you store the enzyme without significant loss of activity? All of these questions should be addressed in order to have adequate control of the enzymes. In the longer term, knowledge of the relationship between the enzyme activity and the characteristics of the process and finished product can help greatly in diagnosing and providing a remedy for problems that may be encountered.

In This Chapter:

Basic Principles

Types of Assay Methods
  Spectrophotometric tests
  Viscometric tests
  pH
  Fluorescence
  ELISA
  Comparison of test results

Specifications

Basic Principles

An enzyme is normally assayed by measuring either the disappearance of substrate or the formation of products under very well-defined conditions. Enzymes with the same activity at a specified quantity should consistently produce the same catalytic activity from one assay to the next. It is most important that the operational parameters such as pH, identity and molarity of the buffer system, reaction time, and the temperature at which the assay is performed be defined and followed closely. All glassware used should be scrupulously cleaned and rinsed with deionized water.

It is also very important to consider the comparative concentration of enzyme and substrate used in any assay procedure. The rate of re-

Specification—Written requirements for formulation, packaging, shipment, and any other relevant characteristics of an ingredient.

action should be a function of enzyme concentration only and be independent of substrate concentration. To achieve this, the substrate concentration must be much greater than the enzyme concentration.

Maintaining a very high level of substrate relative to enzyme ensures that virtually all the enzyme is bound to the substrate so that the reaction proceeds at its maximum rate and depends only on the concentration of enzyme.

Commercially available enzyme preparations are usually derived from bacterial or fungal fermentation and are most often mixtures of enzymes. After the microbes have been allowed to ferment and produce the enzymes as products of their metabolism, purifying the enzymes can be very expensive. To make the preparation affordable, only preliminary purification is usually done. The practical result of this is that the preparation may contain many different enzymes, including both exo- and endoproteases, exo- and endoamylases, and exo- and endopentosanases as well as cellulase and others. This diversity of activities may or may not be evident to either the manufacturer or the end-user. It is not unusual for a manufacturer to market a particular preparation as, for example, an amylase enzyme, when in fact it also has other types of activity present. The supplier's efforts may be focused on making sure that the stated amylase activity is predominant and always the same. The other activities present may not be standardized or even monitored at all. This makes it very important that you understand the enzyme functionality required in your process and be sure that the preparation you use has a verifiably constant level of the relevant enzyme(s). In addition, you should determine whether the preparation contains other enzymes that could affect your process. If differences in processing occur, the manufacturer will verify that its tests show no difference in enzyme activity levels; however, the enzyme causing the processing problems may not be the enzyme activity being monitored by the manufacturer.

Types of Assay Methods

Many different methods are available for the analysis of food enzymes. These include techniques for measuring viscosity, pH, conductivity, specific ions, fluorescence, and absorbance. Spectrophotometric and viscometric assays have historically been used in the food industry, and the equipment required is commonly available in food laboratories. A *spectrophotometric* method uses light to measure changes caused by the enzyme. A *viscometric* assay measures the change in viscosity or flow properties of a liquid. Emphasis on these two techniques is not meant to imply that these methods are preferred for any particular assay. The optimal assay technique should be determined by considering the equipment, expertise, and resources available at a particular facility.

To properly analyze an enzyme, a basic understanding of how the enzyme affects the substrate material is necessary. For example, an

Spectrophotometric—A type of analytical method that measures change in the interaction of a compound with light.

Viscometric—A type of analytical method that measures the change in flow properties of a liquid.

endoenzyme is likely to have a very significant effect on the viscosity of a polymer solution, whereas an exoenzyme has relatively little effect. So, for measuring the activity of an exoenzyme, a viscometric assay would not be a good choice.

SPECTROPHOTOMETRIC TESTS

Spectrophotometric testing methods have always been popular because they tend to be relatively uncomplicated and usually require only basic instrumentation. This assay technique is probably used more often than viscometric methods for many applications. A *colorimetric* test is a spectrophotometric analysis that uses light in the visible color range. One of the original methods for colorimetric α-amylase determination is the Wohlgemuth or SKB method (1), which relies on the ability of starch to form a colored complex with iodine. A normal, intact starch polymer forms a deep blue complex with iodine, but as the starch is degraded by α-amylase into smaller dextrins, the color turns to a red-brown. The length of time required to go from deep blue to red-brown is a measure of the amount of enzyme activity. The shorter the time involved, the greater the amount of α-amylase activity that must have been present.

A spectrophotometric method that is widely used in the enzyme-manufacturing industry for determining proteolytic activity is the Ayre-Anderson method (2). This method, which has been modified many times, is also referred to as the hemoglobin method because hemoglobin is used as the substrate. The enzyme hydrolyzes the hemoglobin, producing amino acids and small peptides. Trichloroacetic acid is added after a suitable reaction time to precipitate the large peptides and unhydrolyzed protein. The mixture is filtered to separate the soluble amino acids and peptides from the precipitated protein, and the absorbance of the resulting solution is measured at a wavelength of about 280 nm. This wavelength is used because it interacts strongly with certain amino acids (tyrosine and tryptophan) that are expected to be in the solution. Higher absorbance values indicate more proteolytic activity.

More recently, assays have been developed using substrate material chemically linked to a colored dye. A method utilizing this idea was developed in 1979 as an assay for α-amylase for use in the cereal industry (3). This test uses a starch substrate chemically linked to an organic dye. The colored starch substrate does not dissolve in water but does absorb water and swell so that it becomes available to the action of the enzyme. The enzyme hydrolyzes the starch, producing smaller dextrins that contain the dye and are water soluble. The result of the enzyme action is a colored solution. The darker the color of the solution, the more α-amylase that is present. This method has been developed as an official AACC method (4). The technique has also been applied to pentosanases, β-glucanases, and other carbohydrate-degrading enzymes (5). The basic approach could also be used to colorimetrically analyze protease activity.

Colorimetric—Describing an analytical method in which the interaction of the analyte with light in the visible range is measured, usually as the absorbance of that light.

VISCOMETRIC TESTS

An example of a viscometric enzyme assay is the use of the amylograph to measure α-amylase activity in flour (6). In this assay, a flour-buffer slurry is heated at a constant rate. As the temperature increases, the starch begins to gelatinize, and the mixture becomes more viscous. If the flour contains very little α-amylase activity, the slurry becomes quite viscous. The presence of α-amylase in the flour reduces the maximum viscosity of the slurry due to hydrolysis of the gelatinized starch.

One of the drawbacks to viscometric measurements is that they are usually not related to the enzyme activity in a linear fashion, i.e., a plot of enzyme activity versus viscosity would not produce a straight line (see Fig. 3-1). This is because the same unit of enzyme activity does not have the same effect on all sizes of a polymer. The action of α-amylase on a slurry containing very large starch polymers produces an immediate and large decrease in the viscosity. But, if the solution contains dextrins (glucose polymers about 25–35 glucose units long), the initial viscosity is low, and the same amount of α-amylase does not have as much effect.

pH

Several other assay techniques are useful in the determination of food enzyme activity. The pH meter can be a very useful tool for a number of enzyme assays. The pH-stat (i.e., stationary pH) is the method of choice when the reaction to be monitored involves the continual liberation or uptake of protons over time, although a simple pH meter can also be used. The pH-stat automatically measures the quantity of acid or base required to keep the pH constant. Glucose oxidase activity can be measured by this method because a proton is liberated during the reaction, or an oxygen electrode can be used to measure the amount of oxygen taken up in the reaction. The pH meter or the pH-stat can also be used with specific ion electrodes to measure reactions involving metal ions. Many reduction/oxidation reactions can be measured with a platinum electrode.

FLUORESCENCE

Fluorescence measurement is a technique closely related to spectrophotometric methods. However, instead of measuring the absorbance of light, the fluorescence method measures light given off by certain types of molecules under specific conditions. This technique is usually significantly more sensitive than spectrophotometric methods, allowing the determination of very small amounts of analyte. Its drawbacks are that the instrumentation is not generally found in food laboratories because of the sensitivity, cleanliness, care, and skill required to achieve reproducible data.

ELISA

An additional technique, enzyme-linked immunosorbent assay (ELISA), has emerged recently and deserves some mention. This technique is based on the interaction between antigens and antibodies. In living systems, antigens trigger production of antibodies, which are produced by immune cells and are able to bind with specific antigens. In ELISA, antigens are attached to a solid support such as a glass test tube or a plastic dish. The antigen reacts specifically with antibodies bound to the enzyme to be assayed. The result is that the enzyme becomes attached to the solid support but is still capable of performing its specific chemical reaction. This technique has been used to measure enzyme activity as well as to check for the presence of a specific enzyme (e.g., to help identify specific microorganisms). Such tests have been used primarily in medical and microbiological applications. However, kits for the food industry are being developed for detection of food additives, fungal and bacterial contamination, antinutritional factors, pesticide residues, and hormones (7). It seems likely that, before long, direct food-enzyme applications may also be developed.

Kits containing prepackaged reagents for the assay of enzymes are available commercially. However, most of the kits are oriented more toward medical diagnostic testing than toward analysis of enzymes of interest to the food industry.

COMPARISON OF TEST RESULTS

Comparing the results from different assay techniques can sometimes be complicated. For example, the results from viscometric testing methods often do not correlate well with those of some spectrophotometric methods. Data resulting from the amylograph method for α-amylase are usually not linear with enzyme concentration; however, results from the SKB method are generally linear over a limited range. Since both methods respond to the *depolymerization* of the starch polymer, one might expect some correspondence between them. Correspondence does exist. However, to analyze these data on the same plot, the nonlinear amylograph *peak viscosity number* must be converted to the *mobility number*, which is the reciprocal of peak viscosity (i.e., 1/peak viscosity) and is linear with enzyme concentration over a limited range. The amylograph mobility number can then be plotted with and properly compared to the SKB number over the range in which the data are linear.

A different situation is presented when comparing, for example, a viscometric protease assay such as the method developed by Northrop (8) with the Ayre-Anderson spectrophotometric method. While not widely used elsewhere in the food industry, the Ayre-Anderson method and variants thereof are commonly performed as standard tests by most enzyme manufacturers and suppliers. There are several major concerns with this method. First, if the extent of

Depolymerization—Breakdown of a polymer.

Peak viscosity number—A number representing the highest value of viscosity attained during an amylograph test.

Mobility number—A number corresponding to the reciprocal of the peak viscosity number, that is, the value represented by (1/peak viscosity number).

proteolysis is limited such that an endoprotease produces only a small number of relatively large fragments from one large protein, these fragments would likely be lost in the precipitation step and thus not measured. Second, hemoglobin, the substrate, has a tyrosine residue only one amino acid from its carboxy terminus, which means that an exoprotease such as carboxypeptidase releases large amounts of tyrosine into solution (tyrosine stays in solution during the precipitation step). However, the size of the hemoglobin molecule is not changed significantly. This is also the case in a variation of the test in which α-casein is used as substrate (α-casein has a tryptophan residue at the carboxy terminus). The resulting release of amino acid results in a large and rapid increase in the absorbance value without a corresponding effect on the size of the protein or the viscosity of the solution. Thus, the Ayre-Anderson test is very sensitive to exoprotease activity but relatively insensitive in detecting endoprotease activity. The viscometric tests, however, are not sensitive to exoproteases—they respond much more to endoprotease activity. Thus, these two tests do not correlate well because they do not measure the same kind of enzymatic activity.

It is important to understand the implications of this. Let's say that you are looking at some data provided by the manufacturer of a new enzyme. It shows the effect of several enzymes on a protein substrate (Fig. 4-1) using the Ayre-Anderson method. The activity may be expressed as "absorbance at 280 nm" or perhaps as "protein solubilized." Enzyme A shows a large increase in the protein solubilized; enzyme B looks less effective; and enzyme C doesn't appear to be very effective at all. The question you should be asking is "effective at doing what?" If you want the proteolytic enzyme to produce lots of amino acids but have no effect on the viscosity, the enzyme represented by curve A may be the most effective for that application (assuming that it is equally effective in hydrolyzing the protein substrate in your product). But if you need a change in viscosity, these data give little information on which enzyme would be most effective.

In many instances, enzyme manufacturers use different tests to measure the activity of enzyme preparations. For example, protease tests results can be in Northrop units (NU/g), tyrosine units (TU/g), hemoglobin units (HU), or in other units such as NPU/g or MCU/g. It is very difficult for the end-user to determine the correspondence (if any) between these units. Be sure you understand what is being measured by these assays and how they relate to the effects you require in your production process.

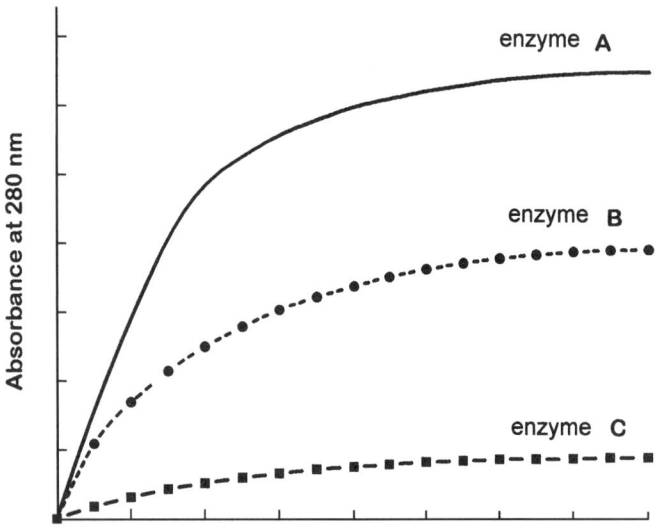

Fig. 4-1. Absorbance data for three protease enzymes as measured with the Ayre-Anderson type of assay.

Specifications

Another aspect of a quality control program is specifications. The number and type will vary depending on the particular requirements, but the following factors should be considered when defining the specifications for any enzyme you need.

Definition. Define the product as completely as possible. If the enzyme is of microbiological origin, specify precisely which organism is used and the particular enzyme(s) allowed to be present in the preparation, the acceptable range of enzymatic activity, and the method for measuring that activity. Any changes in the source of the enzyme, the enzymes in the preparation, or their activity level should require prior approval from the end-user.

Composition. Describe the physical appearance of the material and define any carriers and/or preservatives that may be present.

Legality. Require that the preparation meet all applicable laws, regulations, and provisions of the federal Food, Drug and Cosmetic Acts and any other applicable state and local requirements, including any Kosher requirements.

Packaging. Define the method and material used to package the enzyme.

Shipment. Define the method of shipment and any specific requirements related to shipment (clean conveyance, refrigeration, etc.).

Compliance. Require that acceptance of any shipment be contingent upon compliance with all specifications.

Labeling. Require that lot numbers and corresponding quantities in each lot be noted on packaging and on the bill of lading. Each shipment of enzyme should have a numerical identifier indicating where and when it was prepared.

Size. Define the size of each lot of shipped enzyme and the size of each package within that lot.

Defining these and any other applicable specifications will help to ensure a consistent supply of enzymes.

Appendix D gives primary references for specific enzyme assays. Each assay is briefly described in terms of the basic equipment requirements as well as the basic principle involved.

References

1. Sandstedt, R. M., Kneen, E., and Blish, M. J. 1939. Cereal Chemistry 16:712.
2. Ayre, C. H., and Anderson, J. A. 1939. Canadian Journal of Research 17C:239.
3. Barnes, W. C., and Blakeney, A. B. 1974. Staerke 26:193.
4. American Association of Cereal Chemists. 1995. *Approved Methods of the AACC*, 9th ed. Method 22-06, Cereal α-Amylase. The Association, St. Paul, MN.
5. McCleary, B. V. 1991. Chemistry in Australia 58:398.

6. American Association of Cereal Chemists. 1995. *Approved Methods of the AACC*, 9th ed. Method 22-10, Diastatic Activity of Flour, with the Amylograph. The Association, St. Paul, MN.
7. Allen, J. C., and Smith, C. J. 1987. Trends in Biotechnology 5(7):193.
8. Northrop, J. H., and Hussey, R. G. 1923. Journal of General Physiology 5:353.

Supplemental Reading

1. Aurand, L. W., Woods, A. E., and Wells, M. R. 1987. *Food Composition and Analysis*, AVI, Westport, CT.
2. Pomeranz, Y., and Meloan, C. E. 1978. *Food Analysis: Theory and Practice*. AVI, Westport, CT.
3. Stauffer, C. E. 1989. *Enzyme Assays for Food Scientists*, AVI, Westport, CT.
4. McCleary, B. V. 1991. Chemistry in Australia 58:398.
5. Wilson, R. H., Ed. 1994. *Spectroscopic Techniques for Food Analysis*. VHC Publishers, Inc., New York.

Application of Enzymes to Baked Products

Successful modification of either the baking process and/or the characteristics of finished cereal grain products must be based on an understanding of the basic components of wheat flour and the effects resulting from their enzymatic modification. A normal hard wheat flour (of about 70% extraction rate) is composed of approximately 82% starch, 12.5% protein, 3.5% fiber, 1.5% lipid, and 0.5% ash (minerals). Soft wheat flours have somewhat less protein (about 8–10%) and correspondingly more starch. The best candidates for enzymatic modification are the starch, protein, and fiber fractions.

In general, several effects can be expected as a result of specific enzymatic modification of the flour components. Keeping in mind that commercial enzymes contain multiple forms of enzymatic activity, one can expect that amylases will produce additional fermentable sugars and will alter the dough consistency, making it softer. Proteases will also soften the dough as the result of gluten protein modification and can produce enough new amino groups to affect product color and flavor through the Maillard reaction. Pentosanases will significantly affect dough *rheology* and, depending on the moisture content and type of product produced, can adversely affect product texture.

In This Chapter:

Sources of Amylases

Bread
 Dough
 Product Characteristics

Low-Moisture Products

Troubleshooting

Sources of Amylases

A normal flour, milled from sound wheat, contains significant amounts of β-amylase but little or no α-amylase. β-Amylase can produce some maltose to aid fermentation without the presence of α-amylase, but the amount produced is relatively small, so α-amylase must be added to sound flour from an outside source.

Malted barley has traditionally served as the primary source of added α-amylase. This was the practice years ago when bread formulas were quite lean and contained relatively little sugar. Today, the primary reason for adding amylases to baked products is to improve the processing conditions as well as the overall quality of the baked product. Although barley malt is still used, supplementation with fungal and, to a lesser degree, bacterial amylases has become more common.

The different sources of amylase activity are not equivalent since the amylase enzymes produced from these sources have different

Rheology—The properties of a material that relate to its deformation and flow characteristics.

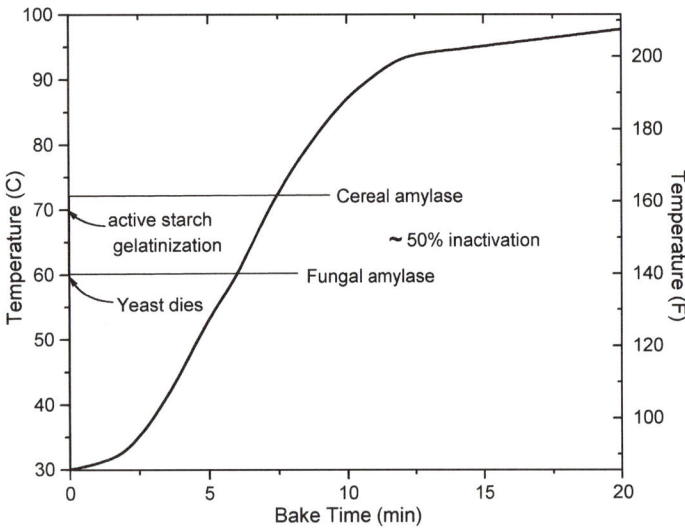

Fig. 5-1. Thermal time course during bread baking, showing relevant temperatures for some formula components. Enzyme temperatures shown for 50% inactivation.

Sprouting (of wheat)— A condition in which high moisture promotes premature growth of the kernel while in the plant head. This leads to increased enzymatic activity in flour milled from such kernels, which may be detrimental to certain baked products.

properties. While most amylases operate best under relatively acidic conditions (pH 5.0–6.5), the temperature ranges over which they are able to operate are quite different.

Fungal amylases tend to be the most susceptible to heat, being inactivated at about 60°C (140°F). Thus, fungal amylases are thought to be relatively safe to use in breads and other high-moisture products since they are inactivated relatively early in the baking process before significant starch gelatinization.

Cereal amylases remain active until the temperature reaches about 70°C (158°F). They can cause some additional starch hydrolysis after the starch has gelatinized. Thus, cereal amylases, whether added to the flour (as a supplement) or native (due to *sprouting*), can produce excessive hydrolysis in high-moisture bread products if too much is present in the formulation.

Bacterial amylases are generally the most heat stable, some remaining active all the way through the bread-baking cycle (Fig. 5-1). They must be considered the most problematic in bread production since they can continue to hydrolyze the starch throughout the baking process. In fact, some bacterial amylases are capable of continuing to work even after baking and packaging and while sitting on the grocery shelf.

Excessive starch hydrolysis results in poor loaf characteristics, including low loaf volume and dense, gummy crumb texture. Because of the difficulty in controlling their activity, bacterial amylases have seen only limited use in higher-moisture product applications, except in the area of shelf-life extension.

The heat stability of bacterial α-amylases is of less concern in products such as cookies and crackers, which have a higher baking temperature and contain little water.

Bread

DOUGH

Hydrolysis of starch. During the dough phase of production, amylases can hydrolyze, to an appreciable degree, only that fraction of the starch that has been damaged in the milling process. Therefore, in the dough stage, the amount of substrate available to amylases is limited by the extent of the starch damage, which varies with the type of wheat from which the flour was milled. In general, hard wheats sustain more starch damage (about 6–8%) than soft wheats (3–5%).

When damaged starch is hydrolyzed by amylases, water that was absorbed by that starch is released to the dough system, making the dough softer. The additional water can also help to hydrate the protein fraction, aiding the formation of gluten.

During baking, starch gelatinizes, making it more susceptible to the action of amylases. Gelatinization is the process in which the intact starch granule absorbs water, swells, and eventually breaks down, releasing the amylose contents of the granule. This process requires both sufficient heat and water. In the presence of excess water, starch gelatinizes at approximately 60°C (140°F). As the amount of water available decreases, the temperature at which gelatinization occurs increases (Fig. 5-2).

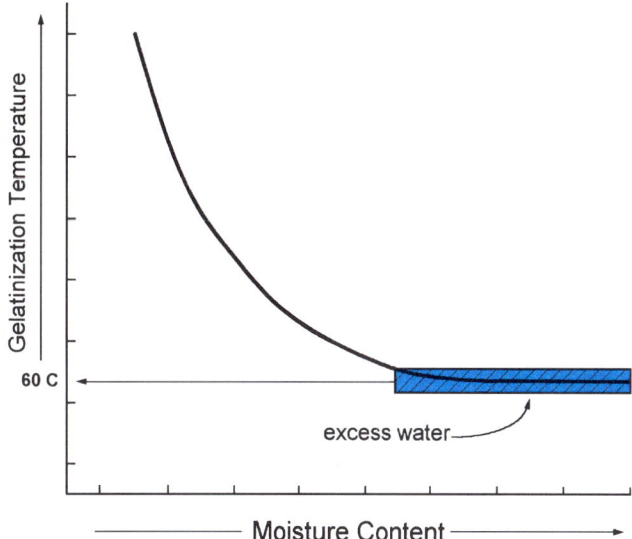

Fig. 5-2. Starch gelatinization behavior as a function of temperature and moisture content.

Gluten development. The protein component makes up a relatively small percentage of the flour compared to starch, but it is responsible for the formation of gluten, the unique viscoelastic protein network peculiar to wheat. Gluten is formed when its component flour proteins, glutenin and gliadin, are mixed together with sufficient energy in the presence of water.

Gluten is both viscous and elastic. That is, it can be stretched without breaking and also tends to pull back toward its original shape. This property is what gives wheat flour doughs their unique consistency. Without gluten, white pan breads, buns, and rolls would not be possible. Because gluten protein is primarily responsible for the mixing and sheeting behavior of dough, it is an obvious candidate for enzymatic modification by proteases.

Supplemention with proteases helps to break down the gluten protein so that the dough is softer and more extensible. The action of the proteases and the resulting more extensible dough have several effects on processing:

1. Reduced mixing time since there is less resistance to mixing.
2. Improved flow characteristics of the dough, easing the process of pan filling for rolls and buns.
3. Improved machining properties. The modern high-speed processing of doughs requires that they be extensible, forming even, smooth sheets without excessive elasticity. Protease treatment substantially lessens the machining difficulties experienced with very tight (bucky) doughs.
4. Improved gas retention. The increased extensibility and pliability of the gluten film helps it to better retain the gas that evolves in the system during processing.

While most commercially available proteases contain relatively nonspecific endo- and exoproteolytic activity, the units of activity will have been determined on nongluten substrate protein. This can make it difficult to choose the most efficient enzyme to hydrolyze the gluten protein in a particular process. The ability of any protease to effectively hydrolyze the gluten must be determined by experimentation in the appropriate formulation and processing scheme. Because the majority of the effects desired on wheat doughs involve altering the viscosity or rheology of the dough, an endoprotease is more effective than an exoprotease.

Lipoxygenase exhibits a gluten strengthening effect and increases the mixing tolerance of a dough. It also oxidizes carotenoid and chlorophyll pigments in flour to their colorless form, which results in a bleaching action to whiten flour. The enzyme is added to flour, usually as a soy flour supplement. This must be done carefully however, since the type 2 lipoxygenase is also able to oxidize fatty acids. Thus, while achieving a whiter flour, one risks a greater likelihood of developing rancidity.

Dough consistency. Doughs containing whole wheat flour and/or rye flours contain pentosans and other carbohydrates that bind additional amounts of water. It has been estimated that the approximately 4% pentosan in a wheat flour is responsible for absorbing about 20–25% of the formula water. As a result, when dough is made with such flours, more formula water is usually added to achieve the proper dough consistency.

Higher pentosan contents (about 9%) can make the dough drier and tighter because so much of the water is absorbed by the pentosan fraction. The resulting requirement for additional water can be substantially decreased by the use of pentosanase enzymes. The enzyme treatment helps to free some water, making the dough less viscous and more extensible.

Commercial pentosanase preparations usually contain a number of other enzymes, including cellulases, β-glucanases, and proteases in addition to the pentosanase. The nonstarch carbohydrates that are the substrates for these enzymes are hydrolyzed to such an extent that they lose the water that was absorbed during the forming of the dough. This hydrolysis can result in rapid and drastic effects on the dough rheology. The primary effect is a significant softening of the dough due to the release of pentosan-bound water. This water is released to the dough, where it is then available to help hydrate the starch and protein fractions and may result in more gluten formation during mixing.

While pentosanase activity produces a softer dough, if the side activities are minimized, the dough should not be sticky and will usually remain machinable. That is the case in high-moisture products such as breads and rolls as well as in crackers and cookies. The dough is quite tolerant to the amount of pentosanase added.

Dough improvement/oxidizing. Interest has increased recently in finding an enzyme to replace chemical oxidizers such as bromate.

Thus far, no single enzyme has been identified that can replace the addition of oxidizing agents. A number of companies have introduced enzyme preparations containing multiple forms of enzyme activity, including amylases, proteases, and several *hemicellulases* in combination with ascorbic acid. The product is claimed to produce dough improvement properties as well as to have the potential for bromate replacement. It is not clear at this time how well these products function in a variety of baked goods, nor is it clear which enzymes are producing the oxidant effect.

Levels of enzyme activity. Doughs prepared with insufficient enzyme activity tend to be tight and overly elastic. They are difficult to machine because they resist being molded and shaped. They may not fill the shape of the pan because of their resistance to flow. The sheeted dough shrinks back too far, increasing the individual piece weight.

The effect of too much enzyme activity in the dough stage is usually observed as a slack, sometimes sticky dough that can be difficult to handle. During proofing, the dough may rise faster due to additional fermentable sugars and weakening of the gluten structure. The doughs are likely to produce machining problems (during dividing, rounding, molding, and/or sheeting) due to their slackness, loss of elasticity, and possible stickiness. The dough has less elasticity, although it may continue to be extensible.

A formulation component that can affect enzyme activity is the fat or oil. At high levels, triacylglycerols can coat the flour particles, making them less susceptible to hydration and subsequent enzymatic hydrolysis. In general, oils are more effective in coating the flour particles than solid fats are. If, in some applications, it is necessary to minimize enzymatic hydrolysis, the flour can be mixed with oil before adding the water to a formulation in order to retard the activity of any enzymes present.

PRODUCT CHARACTERISTICS

Texture. In a typical high-moisture system, hydrolysis of the starch by amylase provides nutrients for the yeast to metabolize, generating CO_2, which contributes to the desired open crumb structure. The soluble dextrins released help to soften the crumb. Protease treatment improves crumb characteristics, and the improved flow properties of the dough contribute to better, more uniform product shape. Pentosanase treatment may provide some improvement in crumb texture, particularly when the bread is formulated with whole wheat or rye flour, which has significantly more pentosan material.

Insufficient protease activity results in a stronger gluten network and consequently in a coarser crumb grain and texture as well as decreased volume.

Excessive enzyme activity produces poor crumb grain and texture. The crumb is denser than desired and lacks the fine, even grain of a properly baked loaf. The texture may be too soft, sometimes "gummy." This retention of moisture in the crumb results from the

Hemicellulase—Any of several enzymes, including pentosanase, glucanase, and cellulase, that hydrolyze nonstarch polysaccharides.

production of too many dextrins from the starch and the loss of gluten structure (with a resulting increase in water-soluble peptides).

Color and flavor. Development of the crust color and the product flavor result from the Maillard reactions that take place during processing. Because the Maillard reaction requires both a reducing sugar and an available amino group, both amylases and proteases can contribute by making these compounds available.

When enzyme activity is too low, fewer sugars and amino acids/peptides are released; the resulting crust color is lighter than normal, and the flavor could be perceived as bland.

Excess enzyme activity affects the color and flavor of the product, usually perceived as an overly dark, hard crust with the presence of "off notes" in the flavor profile.

Loaf volume. In high-moisture systems such as breads and rolls, adding amylases to the formulation results in increases in the loaf volume of the baked product. It is not clear at this time whether this is an effect of the amylases, a side activity of proteases, or the result of the combination of enzymes. Protease treatment makes the dough more extensible so that it is better able to retain CO_2. Thus, the greater loaf volume attributed to amylase enzymes may well be a function of both amylases and proteases.

If enzyme activity is insufficient, the resistance to flow caused by the stronger gluten network results in decreased loaf volume.

Too much protease activity results in the loss of gluten structure, which decreases the ability of the dough to retain the expanding bubbles of CO_2 as the dough heats. Therefore, although the dough may initially rise to a greater volume, the inability to retain the gas results in collapse of the loaf structure and ultimately a decrease in loaf volume.

Shelf life. Extension of shelf life is a rather specific effect, associated only with bacterial amylases because they can continue to act after baking. The mechanism of this positive effect is not understood precisely but seems to be related to retarding or inhibiting the formation of bonds that otherwise result in bread firming. Current theories suggest that starch recrystallization is inhibited or that bonds between the starch and protein fractions are prevented from forming. It is presumed that this antistaling effect is contributed by oligosaccharides of a certain size range since the distribution of oligosaccharides produced by bacterial amylases is quite different from that resulting from cereal or fungal amylase hydrolysis of starch.

Low-Moisture Products

In low-moisture products such as crackers, cookies, and pizza crust (thin crust), starch gelatinization is limited and contributes less to the development of product structure. Crackers and cookies also require only limited gluten development. For these reasons, release of water by amylase or protease would be detrimental, as additional water could contribute to uneven and excessive cookie spread. Excess starch hydrolysis by amylase is likely to produce a very hard, flinty texture. Excess protease activity causes the piece to be flat and overly tender, leading to excess breakage in the package.

The temperature of the low-moisture baking piece rises very rapidly, reaching higher temperatures than those of breads, so enzymes may remain active for a shorter time. In addition, the relatively low water content tends to slow enzymatic activity.

However, the quality of the finished product is usually enhanced by protease treatment. In low-moisture products, high gluten content can lead to textural properties that are perceived as "tough." Protease treatment leads to a tenderizing effect in these products.

Pentosanase treatment can also benefit cookies and crackers. The hydrolysis of the pentosan network has several consequences. First, it releases the pentosan-bound water. If this effect has been compensated for in the formula water (i.e., less formula water), this treatment can result in more consistent spread so that the packaging lines operate more efficiently. Second, the released water is easier to bake out since it is no longer bound in the pentosan network. Thus, the desired moisture content can be achieved in a shorter bake time, saving energy and lowering production costs.

In low-moisture products, machining problems in sheeting the dough sometimes lead to increases in piece thickness and weight which, in turn, alter the baking characteristics. The thicker piece retains more moisture. This can lead to increased starch gelatinization, producing a harder texture in the product after baking. In such doughs, added pentosanases and amylases can release bound water to help produce a softer, more pliable dough.

A particular problem arises when a bacterial, rather than a yeast, ferment is used, as is common in the production of sour breads and soda crackers. The pH at the beginning of the fermentation is usually slightly acidic, about 6.5. As the fermentation proceeds, organic acids are produced (primarily lactic and acetic), which decrease the pH of the media over time. The final pH can be quite acidic, reaching levels near 3.5. This change in pH is sufficient to significantly alter the activity of many commonly used enzymes. The rate at which the bacteria grow and produce acids determines how fast the pH of the fermentation media drops. You must determine whether it is better, in your system, to add a neutral protease, which acts before the pH drops; an acid protease, active after the pH drops; or perhaps a combination that will provide more or less continuous hydrolysis throughout the fermentation. Remember also that wheat contains a naturally occurring acid protease that can hydrolyze gluten protein when the pH of the dough drops below 4.0. This is an important contributor to altering dough rheology in processes involving bacterial fermentation.

In low/no-fat cracker products, there is less fat to form a barrier to the entry of water into the starch granule. In addition, less fat results in a much tighter dough, requiring the addition of more formula water to form a machinable dough.

Troubleshooting

The following table lists common problems, causes that relate directly to enzymes, and changes in formulation or processing that may help to remedy the problems.

BREAD		
Symptoms	**Causes**	**Changes to Make**
Tight or "bucky" dough	Insufficient protein modification	Increase protease enzyme activity. Check whether temperature of enzyme makeup water is too high.
	Flour contains too much pentosan	Add pentosanase activity.
	Temperature of dough is too low	Increase temperature of formula water.
Loose, slack dough	Excessive modification of protein and/or starch	Decrease enzyme activity added to formula. Lower temperature of makeup water.
Poor loaf volume	Insufficient gluten protein modification	Increase protease activity. Increase temperature of formula water.
	Insufficient starch modification (particularly in low-sugar formulation)	Increase α-amylase activity. Increase temperature of formula water. Add sugar to formulation, if possible.
	Excessive modification of starch and/or protein	Decrease enzyme activity added to formula. Decrease temperature of formula water. Decrease proof time.
Crumb too soft, "gummy"	Excessive modification of protein and/or starch	Decrease enzyme activity added to formula. Decrease temperature of formula water. Decrease proof time.
	Use of heat-stable bacterial amylase	Use amylase activity from fungal or cereal sources rather than bacterial sources.
Crust color too light, flavor bland	Insufficient amylase/protease activity	Increase amount of enzyme activity added to formula. Increase temperature of formula water. Increase proof time. Add nonfat dry milk to formulation.
Crust color too dark	Excessive amylase/protease activity	Decrease amount of enzyme activity added to formula. Decrease temperature of formula water. Decrease proof time.
Thick-walled cell structure	Excessive protease/amylase activity	Decrease amount of enzyme activity added to formula. Decrease temperature of formula water. Decrease proof time.
Sidewall weakness ("keyhole")	Excessive starch breakdown	Decrease amylase activity. Select appropriate amylase type.
Capped top crust	Excessive starch modification	Decrease amylase activity.

CRACKERS

Symptoms	Causes	Changes to Make
Piece weight heavy, poor product shape	Insufficient protein modification	Increase protease activity.
Piece weight light, poor product shape	Excess protein/starch modification	Decrease added enzyme activity. Decrease temperature of formula water.
Inconsistent product shape	Too much elasticity in dough	Add protease activity.
Tough texture	Excess gluten formation	Increase protease activity. Decrease mixing time. Increase temperature of formula water.
Texture hard, "flinty"	Excess starch hydrolysis	Add fungal amylase activity in dough stage. Decrease formula water. Add pentosanase activity.

PIZZA CRUSTS (CRACKER OR THIN CRUST)

Symptoms	Causes	Changes to Make
Poor shape	Too much dough elasticity	Add protease activity.
Tight, bucky dough	Insufficient protein modification	Increase protease enzyme activity. Check whether temperature of enzyme makeup water is too high.
	Flour contains too much pentosan	Add pentosanase activity.
	Temperature of dough is too low	Increase temperature of formula water.
Loose, slack dough	Excessive modification of protein and/or starch	Decrease amount of added enzyme activity. Lower temperature of makeup water.
Inconsistent piece weight	Tight dough	Increase protease enzyme activity. Check whether temperature of enzyme makeup water is too high. Add pentosanase activity. Increase temperature of formula water.

Supplemental Reading

1. Pyler, E. J. 1988. *Baking Science and Technology*. Sosland Publishing Co., Merriam, KS.
2. Kruger, J. E., Lineback, D., and Stauffer, C. E., Eds. 1987. *Enzymes and Their Role in Cereal Technology,* 2nd ed. American Association of Cereal Chemists, St. Paul, MN.
3. Pomeranz, Y., Ed. 1988. *Wheat: Chemistry and Technology,* 3rd ed. Vols. I and II. American Association of Cereal Chemists, St. Paul, MN.
4. Fox, P. F., Ed. 1991. *Food Enzymology*. Elsevier Applied Science, London.
5. Hoseney, R. C. 1994. *Principles of Cereal Science and Technology,* 2nd ed. American Association of Cereal Chemists, St. Paul, MN.

Enzyme Applications for Beverages

CHAPTER 6

Beer and Ethanol Production

Beer has been brewed for over 5,000 years, one of the first systematic uses of enzymes in the production of a food product. Today's brewing practices are technologically more efficient, but the basics of beer making have not changed significantly since ancient times. The method for producing ethanol in distilled spirits and for fuel is similar to the brewing process.

Most beer brewing begins with barley malt, although malted sorghum is used in some beers produced in Africa. Malts serve as the source of both the starch and protein raw materials and the enzymes required to convert these raw materials into the sugars, amino acids, and peptides needed in the brewing process.

The sugars produced are fermented by a *Saccharomyces* yeast strain and converted to ethyl alcohol and carbon dioxide. The yeast uses the amino acids and peptides to produce flavor compounds as well as for its own growth and reproduction.

BARLEY MALTNG AND ENZYME ACTIVITY

Many breweries purchase malt; therefore, the malting process is often separate from the brewing operation. Malting is the controlled germination and subsequent drying of the barley seed, which results in increased activity of the *endogenous enzymes*. Although a number of enzymes are produced, the most important for the brewing process are the α- and β-amylases and the proteinases. These enzymes degrade starch and proteins to sugars, amino acids, and small peptides.

Having the proper level of enzymes is very important to a consistent brewing process. This level, however, is sometimes difficult to maintain over time. Both the type of barley and the malting conditions can produce variations in the amount and type of enzymes present. Another source of inconsistency in the level of enzymes available is the use of grains other than barley malt as sources of carbohydrate.

In the traditional method for brewing beer, only barley malt is used. In Germany this practice is demanded legally by the "Reinheitsgebot" or purity law, passed in 1516. In other areas of the world, however, other grains, in addition to barley, may be added. These adjuncts, which are cheaper or more readily available, provide a supplemental source of carbohydrate (starch) for conversion to sugars and subsequent fermentation by the yeast. Typical adjuncts include coarsely

---

In This Chapter:

Beer and Ethanol Production
 Barley Malting and
  Enzyme Activity
 Role of Enzymes in the
  Brewing Process
 Low-Calorie Beers
 Distilled Products and
  Ethanol Production

Wine Production

Fruit Juice Processing
 Noncitrus Fruit Juice
 Citrus Fruit Juice

Troubleshooting

Endogenous enzymes— Enzymes occurring naturally in the substance (e.g., in a plant).

Exogenous enzymes — Enzymes from another source added to a substance (e.g., microbial enzymes added to plant substances).

Mash—The mixture of malted barley and water constituting the first step in brewing beer.

ground particles (grits) of wheat, corn, rice, or unmalted barley. Maltose syrups or other sugar syrups may also be used.

When adjuncts such as unmalted barley or wheat are used to replace some of the malted barley, the level of α-amylase and protease enzymes is decreased. Therefore, microbial α-amylases and proteases may be added to supplement the malted barley enzymes. Using these *exogenous enzymes* ensures that the enzyme level is sufficient to modify the starch and protein fractions. Almost half the malt can be replaced if the endogenous enzymes are supplemented.

Heating the adjuncts makes them more susceptible to enzyme action. The starch granules of adjuncts such as corn and rice gelatinize at a higher temperature than wheat and barley starch granules. Microbial enzymes are useful with these adjuncts because they tend to be more heat stable than their cereal counterparts. Heating also thickens the material. Microbial amylases may be added to decrease the viscosity for pumping because they are more economical and work at higher temperatures than cereal amylases.

While addition of exogenous enzymes can help to resolve problems related to insufficient malt enzyme activity, excess malt enzyme activity cannot be easily remedied. Malt should be seen as the first line of defense in preventing enzyme-related production problems. This involves having a comprehensive quality control program in which the quality of the malted barley is defined as completely as possible, including specifications defining the type and presence of the enzymes known to be important in the brewing process.

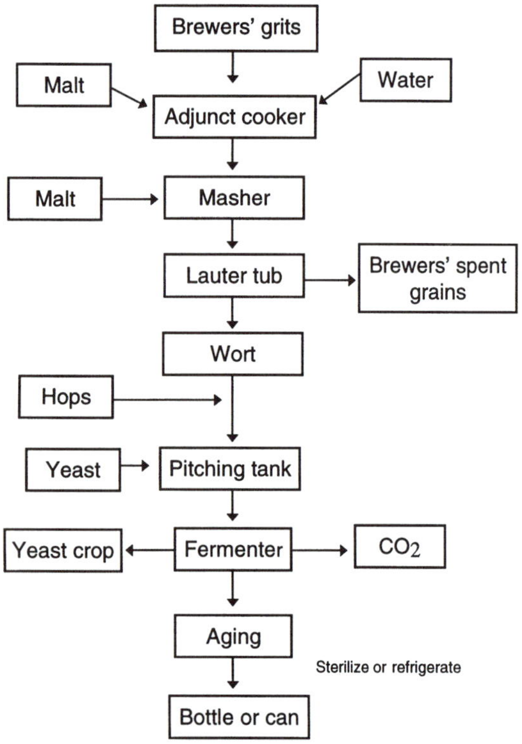

ROLE OF ENZYMES IN THE BREWING PROCESS

Enzymes are used in several of the steps of the brewing process (Fig. 6-1).

Mashing. In the mashing step, the milled barley malt is mixed with water and then the mixture is held at several temperature levels for specified times (Fig. 6-2). These stages vary somewhat among types of beer and manufacturers, but the purpose for the stages is to maximize the effectiveness of the enzymes by holding the water-malt mixture, referred to as the *mash*, at temperatures at which different enzyme activities are optimized. In the mashing process, both the protein and the carbohydrate components are enzymatically modified. The pH is around 5.5, which helps to maximize conversion of starch to fermentable sugars by the amylases and promote breakdown of proteins that would later cause turbidity in the beer. The idea is to create a nutrient-rich medium for the yeast utilized in the subsequent fermentation step.

Generally, mashing begins with a 30-min rest at about 35°C to allow hydration of the malt. Then the

Fig. 6-1. Simplified flow of beer production. (From 1; used by permission)

temperature is increased to 45–50°C and held for about 30 min, allowing the action of proteolytic enzymes that are susceptible to heat. The next step, about 30 min at about 65°C, optimizes β-amylase activity. The mash temperature is finally raised to about 75°C. At this temperature, the starch component will have gelatinized and become available to the action of α- and β-amylases. The mash is held at about 75°C until the majority of the starch has been hydrolyzed to a sufficient degree that it can no longer form the blue amylose-iodine complex. If the iodine test indicates incomplete starch hydrolysis, it is likely that the α-amylase activity was insufficient to accomplish the job.

There are several possible causes for insufficient α-amylase activity. First, the problem could arise from the malt itself, either because it was a poor malting variety of barley or because the malting process was flawed. Second, the mashing process could have been improperly programmed such that the temperature was either too low to completely gelatinize the starch in the time allowed or too high, inactivating the enzyme too early in the process. An immediate remedy would be to add a commercial α-amylase to complete the starch hydrolysis before filtering the mash and proceeding to the fermentation step. In some instances, too fine a particle size or high early mash temperatures result in settling. Heat-stable α-amylase can be added to loosen the mass.

If the starch conversion was not adequately accomplished, protein hydrolysis may also be incomplete. Approximately two-thirds of the protein is made insoluble by the heating process. Aided by the proteolytic enzymes present, the other third of the protein is solubilized. Some adjuncts contribute very little protein; therefore, proteases that utilize the protein more efficiently may be added to make the protein breakdown products available for the yeasts.

Wort. The next step involves filtering or lautering the mash. This step separates the liquid portion, called the *wort*, from the remaining solid residue or spent grain. At this stage, the wort has a *dextrose equivalent* level of about 40 and contains sugars, dextrins, amino acids, and small peptides.

During filtration of the mash, a slow runoff rate is usually an indication that the malt was relatively high in β-glucans and/or low in the β-glucanase enzyme, which hydrolyzes this carbohydrate. Poorer malting barley varieties often have a higher β-glucan level than better malting barley varieties. The higher β-glucan level in the mash leads to higher viscosity and subsequent difficulty in filtering the mash. In some cases, a gelled layer may form that clogs the filter holes. While these malted barleys do contain significant levels of β-glucanase en-

Wort—The liquid remaining after filtering the finished mash.

Dextrose equivalent—A measure, based on the reducing power of glucose, of the percentage of glucosidic bonds in a starch polymer that have been hydrolyzed.

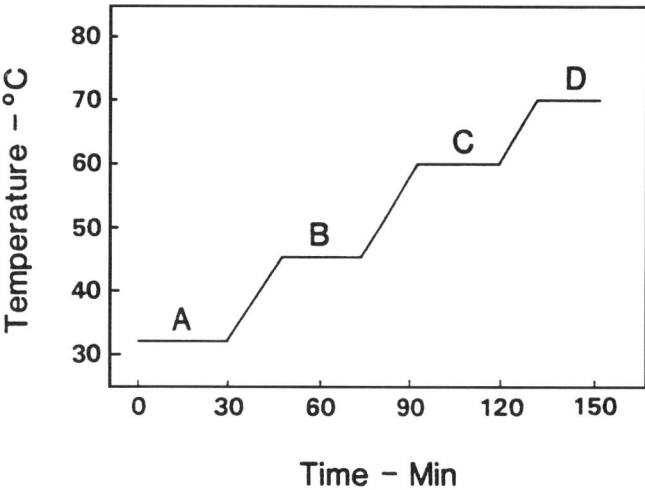

Fig. 6-2. Outline of the mashing process. A, acid rest; B, proteolytic rest; C, β-amylase optimum; D, α-amylase optimum. (Reprinted from 1; used by permission)

zymes, their optimal pH level is about 4.0–4.5, below that of a normal mash, which is about 5.75. These β-glucanases are also rather unstable at the mash temperature, so they are not effective at hydrolyzing the β-glucan material and reducing the viscosity. This is another situation in which the more heat-stable commercial microbial β-glucanase can be of value. Again, the long-term solution is to institute a quality control program eliminating such problem malts from the processing.

After filtering, the wort is boiled with hops (the dried fruit of the plant *Humulus lupulus*, which provides aroma and flavor compounds). This step concentrates the wort, sterilizes the liquid, coagulates protein, and inactivates any remaining enzymes. Many of the flavor components from the hops are volatile and so are lost during boiling. For that reason, it is common to add about half the hops at the beginning of the boiling process and the other half toward the end. The hot wort is cooled and filtered and is then ready for fermentation. Great care must be exercised during this stage to prevent unwanted microorganisms from infecting the now-sterile wort.

Pitching and fermentation. Yeast is then added to the cooled wort, a process called *pitching*. Many individual strains of yeast are used by different brewers. In addition, the type of beer (pilsner, ale, stout, etc.) also dictates the choice of yeast. The identity of the specific strains is considered a trade secret among competing brewhouses.

The process of converting the fermentable sugars to carbon dioxide and alcohol is a rather complex one, with many different enzymes involved. The reader is referred to a biochemistry text that explains in detail the Embden-Meyerhoff-Parnas pathway for alcoholic fermentation (see the supplemental reading at the end of the chapter). The reaction of interest is illustrated here in very simplified terms:

$$\underset{\text{Glucose}}{C_6H_{12}O_6} \rightarrow \underset{\text{Carbon dioxide}}{2\,CO_2} + \underset{\text{Ethyl alcohol}}{2\,CH_3CH_2OH}$$

In the fermentor, the two problems of most concern are manifest as either a slow rate of fermentation or a low alcohol level in the fermented wort. A low fermentation rate implies that the yeast did not have an adequate supply of nutrients to sustain the desired rate of fermentation. This, in turn, suggests that the enzymatic conversion of protein and starch occurring during the mashing process was not adequately accomplished. It is possible to correct this situation by adding commercial exoprotease and α-amylase activity in order to provide the proper degree of hydrolysis and thereby supply more yeast nutrients. A low alcohol level in the fermented wort indicates that the extent of starch conversion was too small to produce the target alcohol level. This may also be corrected by addition of a commercial α-amylase enzyme. The fermentation step is complete when the desired sugar concentration is reached.

Maturation. After fermentation, the "green beer" requires a maturing step for development of flavor and foam quality.

Pitching—Adding the yeast to the wort, initiating fermentation.

During the aging process, some partially hydrolyzed protein tends to aggregate, coming out of solution and forming a hazy suspension. This haze reduces the clarity of the beer and can present difficulties with filtration. Haze may form if mashing is inadequate (either from too high a temperature or low enzyme activity). It may also be observed during or after chill proofing of the beer. The addition of an exoprotease usually will alleviate the condition by further hydrolyzing and solubilizing the offending proteins.

Care must be taken not to overuse proteases because they may break down the proteins so much that foam does not form. Proteins with higher molecular weight may contribute to haze formation, but they are also essential for proper foam formation and retention. Larger proteins are required to act as bridging agents (to stabilize the foam). Excess protease activity hydrolyzes them to smaller peptides no longer capable of forming a stable, long lasting "head" on the beer.

Accumulation of diacetyl is another common problem during aging. Yeast secretes α-acetolactate, which is an intermediate of diacetyl formation. Diacetyl is absorbed by yeast and reduced to 2,3-butanediol, which does not cause off-flavor. However, if no yeast is present, the diacetyl accumulates, producing an undesirable flavor. Two common ways to remove or prevent diacetyl accumulation are to ferment with organisms without acetohydroxy acid synthase so that less diacetyl is produced or to add enzymes that accelerate the breakdown of acetolactate while the yeasts are still present to absorb the diacetyl.

After aging, the beer goes through a final filtration before bottling. Antioxidants such as ascorbic acid may be used to help prevent oxidation in the bottled beer. The oxygen content of the beer is virtually zero after fermentation, but some oxygen may enter the headspace of the containers during packaging. The addition of an oxygen scavenging enzyme to the beer itself could help to remove this headspace oxygen, but an oxygen scavenger incorporated into the packaging (in the bottle cap, for example) may probably be more effective and cost efficient.

LOW-CALORIE BEERS

Breweries in the United States brew beer from a wort with a solids level of about 15%. This is fermented to about 75%, resulting in a beer with an alcohol content of about 6.5%. The beer is then blended with water according to the Food and Drug Administration guidelines defining the percentage of alcohol allowed in beer (3.2–4.5% by volume). This situation is complicated by the fact that each state has its own limits on alcohol content of beer, so the "normal" beer in any state varies in alcohol content within the allowable range.

Under normal conditions, the α- and β-amylase enzymes used in the brewing process cannot break down the α-1,6 bonds found in the amylopectin fraction of the starch. Therefore, after fermentation, remnant carbohydrates composed of pieces of the starch having α-1,6 branch points, as well as some pieces greater than three glucose units long that are not easily hydrolyzed and are not fermentable, remain

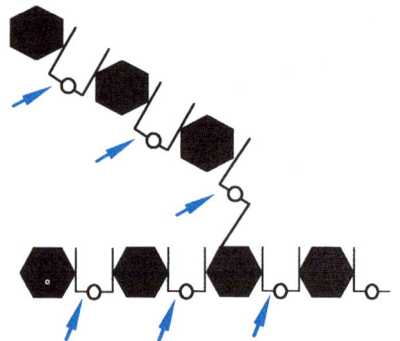

Fig. 6-3. Remnant carbohydrates, which are removed in the production of low-calorie beer.

in solution. An example remnant carbohydrate is shown in Figure 6-3. Since they remain in the finished beer, they contribute to the total caloric value.

Several methods are used to produce low-calorie beer, defined (in the United States) as having 30% fewer calories than a manufacturer's "normal" beer. One method for decreasing the total caloric content is to simply dilute the beer with water. This, however, is not a preferred solution because in order to reduce the total caloric content enough to meet the requirements for low-calorie beer, both alcohol and carbohydrate sources of calories are diluted and the final alcohol content is lower than that of a normal beer. This decrease in both alcohol and carbohydrate negatively affects the flavor and mouthfeel of the beer.

Alternatively, manufacturers can produce low-calorie beers using enzymes to remove the carbohydrate remnants. The enzyme of choice is a glucoamylase from *Aspergillus niger* that can hydrolyze the remnant carbohydrate to glucose because it is less specific than any of the malt amylases. The glucose is then fermented, virtually 100%, to carbon dioxide and alcohol.

The caloric content of glucose, as of most carbohydrates, is about 4 cal/g, and the caloric value for alcohol is about 7 cal/g. Each glucose is fermented to two alcohol molecules and two moles of CO_2. Therefore, if 100 g of glucose (400 cal) is fermented to approximately 50 g of alcohol (350 cal), the decrease in caloric value is only about 12.5%. Thus, simply reducing the carbohydrate to glucose and fermenting to alcohol cannot, by itself, provide the required decrease in caloric content to meet the 30% reduction in calories required to qualify as a low-calorie beer. Further blending with water is still required. The advantage gained by enzymatically hydrolyzing all the remnant carbohydrate to fermentable glucose is that the enzymatically treated 15% wort contains about 10–12% more alcohol but fewer carbohydrates than "normal" beer. This wort can be blended with water to achieve a specific alcohol content. Because the alcohol content of the wort is greater, the final alcohol content of the low-calorie beer is the same as that of "normal" beer, while the caloric contribution from the carbohydrate is sufficiently reduced to meet the 30% requirement. The major reduction in calories has come from diluting the carbohydrate source rather than the alcohol. Because the alcohol content is the same as a "normal" beer, more of the desired flavor and mouthfeel are retained.

DISTILLED PRODUCTS AND ETHANOL PRODUCTION

Ethanol production in distilled spirits or for fuel use is closely related to beer production. Like beer, the carbohydrates are broken down to provide nutrients for yeast during the fermentation step. However, distilled spirits undergo an additional distillation step, and the products generally go through an extended aging cycle, years rather than months. Federal regulations set strict standards defining how each type of liquor is made, labeled, advertised, and sold. These products are unique in that the original ingredients are not present in the final product. The basic categories are brandy, cordials, gin, rum, tequila, vodka, and whiskey (Table 6-1).

Although various types of liquors have a standard of identity for the carbohydrate source, many different sources of carbohydrates are used. For example, bourbon must be made from at least 51% and no more than 79% Indian corn in the fermented grain mash. The carbohydrate source for rum is sugar from cane or its by-products, primarily black strap molasses. Whiskeys generally use malted barley for the source of enzymes, while the grain-neutral spirits use enzymes from a variety of sources.

Ethanol is also produced as a source of fuel. It can be added to gasoline at a level of up to 10% to oxygenate the fuel and help reduce pollution. The production method is the same as for potable ethanol, but there is more concern for having the lowest cost instead of en-

TABLE 6-1. Types of Distilled Spirits

Type	Description
Bourbon	Must be made in the United States from at least 51% and no more than 79% Indian corn in the fermented grain mash; distilled at 160 proof or less; matured in new American white oak barrels that have been charred on the inside, aged for at least two years (though most bourbon is aged at least four years). No blending is permitted; no additives except pure water, which may be added to adjust the proof when bottling.
Brandy	Distilled from fermented grape juice. Caramel may be added to adjust color.
Cordials (liqueurs)	Flavorful distilled spirits containing at least 2.5% sugar by weight or about 1.5 tablespoons/liter. Classified by ingredients and by production method: infusion or maceration, percolation, or distillation with flavoring ingredients.
Gin	Made by mixing spirits with flavoring ingredients, primarily juniper berries.
Rum	From sugar from cane or its by-products, primarily black strap molasses.
Scotch	The primary base grain is malted barley. Distinctive smoky flavor results from the barley malt being dried over a peat fire.
Tequila	Made from sugars from the agave plant.
Vodka	Usually made from grain and not potatoes. Sometimes treated with charcoal so that it will be without distinctive aroma, taste, or color.
Whiskey	
American	A grain mix that is no less than 51% corn; the rest is generally rye and barley malt. Distilled at 160 proof or less and aged at not more than 125 proof in charred, new oak barrels.
Canadian	Rye and corn form the base; can blend whiskies either before barreling and maturing or after the aging period. Used, charred oak barrels can be employed.
Irish	Made from malted and unmalted barley and spirits from other grains.

suring flavor development. Alternative sources of sugars will become available in the future. Production of sugars from biomass is not yet economically feasible.

Wine Production

Fruits and berries are used throughout the world in the food industry. While different in many respects, they share some characteristic common constituents, specifically pectin, arabinoxylans (pentosans), and β-glucans, which can contribute to problems in the processing of these raw materials.

For the most part, the source of wine is grapes. The quality of wine is highly dependent on the quality of the grapes when they are harvested. In addition to variations in the characteristics of different grape varieties, weather conditions can affect the quality and amount of wine the vintner is able to produce. Wet weather can encourage microbial contamination of the grape and cause premature rotting. Dry weather can stunt the growth of the grapes so they are poorly formed and underdeveloped. These problems as well as some others, can be mitigated by the proper application of enzymes, which play an important role in extraction of the juice, subsequent filtration, and wine clarity.

The first step in the production of wine involves pressing the juice out of the grapes. Grapes contain relatively high levels of high molecular weight carbohydrate polymers such as pectin. As the grapes ripen, the protopectin found in the unripe grape is hydrolyzed by native grape pectinases to a softer, more soluble pectin. Unless this process is allowed to continue for an extended time, which would mean allowing the grape to get very overripe, the native grape pectinases cannot degrade the protopectin all the way to its component saccharides. Thus, the partially solubilized pectin absorbs large amounts of the juice of the grape, decreasing the juice yield. The increase in pectin solubility allows some of the pectin to be removed with the juice during pressing, contributing to an increase in juice viscosity. This situation leads to problems during filtration and affects product clarity.

To alleviate these problems, it is common today for manufacturers to add an exogenous commercial pectinase during the mashing, or pressing, operation to hydrolyze the pectin sufficiently to release the absorbed grape juice and reduce pectins levels in the extract, increasing the rate and efficiency of filtering. Use of pectinase enzymes can significantly increase the yield of juice from this first pressing. *Depectinization* of the grape juice is also necessary to ensure a clear juice for subsequent wine production and to prevent the concentrated or stored grape juice from gelling.

Some grapes are harder to process than others because they have more pectin material as well as additional polymeric material in the form of hemicellulosic material, especially arabinoxylans. These grapes, particularly Thompson Seedless and Muscat Gorda varieties,

Depectinization—The breakdown of the pectin polymer with enzymes.

are especially hard to press. Treatment with a mixture of pectinases, xylanases, and cellulases breaks down the carbohydrates and increases juice yield.

β-Glucanase enzymes also find application in winemaking for several reasons. Under certain conditions, the wine grapes can be attacked by a fungus called *Botrytis cinerea*. This fungus, as a part of its own metabolism, produces β-glucans, which pass into the wine. β-Glucans are large carbohydrate polymers that hinder the clarification of the wine by rapidly clogging the filters. Addition of a commercial β-glucanase can hydrolyze the β-glucan polymers down to glucose and small oligosaccharides, which pass through the filters without clogging them.

β-Glucans are also involved in the flavor of the wine. Some flavor components called *monoterpenes* become bound to the glucose molecules making up the β-glucans. These bound flavor components are nonvolatile and so do not contribute to the flavor profile of the wine. Treatment with a β-glucanase enzyme hydrolyzes the β-glucan polymer, releasing the monoterpene. The liberated monoterpene is volatile and adds to the flavor.

In addition to these carbohydrate-degrading enzymes, side activities of proteases are of use in wine production. Although the protein content of grape juice is low, the protein component can contribute to haze in the filtered juice as well as to the formation of foam during fermentation. The protein fragments created by the protease contribute to the fermentation process by being utilized as nutrients for the yeast. A side benefit may be less foaming during the fermentation. The protease treatment may also contribute to greater color stability in red wines by reducing the binding of polymerized tannins to the proteins in the wine.

Monoterpenes—Organic flavor compounds in wine.

Fruit Juice Processing

The problems associated with the processing of fruit juices are similar to those described for wine. Nearly all fruits and berries contain pectic and other carbohydrate materials and, as seen for wine, can cause problems during clarification and filtration.

NONCITRUS FRUIT JUICE

Fruits are generally picked before becoming fully ripe in order to maintain their firmness. In addition to pectic substances, unripe fruits, particularly apples and pears, contain significant amounts of starch. The starch exists in microscopic granules that, like cereal starch granules, are insoluble in cold water. However, heating during processing makes the starch more soluble. In the absence of any enzyme treatment, the starch is not removed by filtration and is not precipitated in subsequent clarification steps. Several days or weeks after clarification, the starch molecules begin to aggregate, forming a starch haze in the juice. The aggregated starch is resistant to enzymatic degradation. This problem can be overcome by adding an α-amylase enzyme during the heating of the pressed juice. As the starch becomes solubilized at about 60°C, the α-amylase degrades the starch into glucose, maltose, and small oligosaccharides, which do not cause subsequent haze formation.

Pectic substances in fruits cause additional processing problems. Probably the largest use of exogenous pectin-degrading enzymes is in the clarification of fruit juice. The raw, pressed juice tends to be rather viscous because of the presence of water-soluble pectin. In addition, the pressed juice contains a persistent cloud of cell wall fragments. The viscosity of the juice makes filtration and removal of this cellular debris difficult. Addition of an exogenous pectinase reduces the viscosity of the fruit juice and causes the suspended cloud particles to aggregate into larger particles that can be removed either by centrifugation or filtration. This is best accomplished using a combination of pectinesterase and an endoacting pectin lyase (EC 4.2.2.10).

The soluble pectin also makes further extraction of juice from the pulp more difficult. Pectinase enzymes can alleviate the poor juice yields by depolymerizing the soluble pectin material, making filtration of the juice easier and more efficient as well as increasing the yield of juice from the pressed fruit pulp.

Another problem associated with the pectin content of fruit juices occurs during the production of fruit juice concentrates. As the fruit juice is concentrated, the water-soluble pectin forms a viscous semi-gelled mass that is very difficult to press out or concentrate further. Treatment with a pectinase enzyme results in a free-flowing liquid that can be filtered and concentrated to a much higher level of dissolved solids.

Arabinoxylans can also contribute to haze formation in fruit juices. As techniques for increasing the recovery of juice from fruits have improved, more of these arabinoxylans have ended up in the juice. The haze resulting from arabinoxylans appears some time after the juice has been filtered and concentrated. The danger of subsequent haze formation can be practically eliminated by the application of commercial pectosanase enzymes. In fact, it is common practice in the industry to use a combination of pectinases and xylanases in the treatment of fruit juice extracts.

Another common component of fruits is cellulose. While cellulose is not water soluble, it does absorb water to a significant extent. This may be important in the production of dried fruit (and vegetable) powders. Cellulases are added during apple juice processing to improve yields and color. The application of suitable cellulases allows for efficient concentration of the fruit extracts by breaking down the cellulose polymer network, releasing large amounts of water and making it easier to evaporate.

CITRUS FRUIT JUICE

Pectic substances are found throughout most citrus fruits, mostly in the peel and the membranes. They present a number of difficulties with respect to citrus fruit processing, especially affecting extractability of the pressed fruit pulp, the viscosity level for effective concentration of the juice, cloud stability, and the extraction of essential citrus oils.

The problems having to do with extracting the pulp and reducing the viscosity of the juice for concentration are similar to those of other fruits and are addressed in the same way as described in the previous section. The issue of cloud stability is peculiar to citrus fruits and to orange juice in particular. Orange juice and other citrus juices are normally not meant to be clear because much of the desirable color and flavor characteristics depend on components in the insoluble cloudy materials of the pressed juice. The problem of *cloud loss* is a function of an endogenous pectin, methylesterase, which deesterifies the pectin to pectic acid. This charged pectic acid interacts with calcium ion (Ca^{++}) in the juice, forming a coagulated complex. On standing, the juice separates from the gellike complex, which settles out as a layer of sediment. This undesirable result of the pectin methylesterase can be overcome by heating the juice to inactivate the enzyme, but heating has detrimental effects on the flavor components of the juice. Thus, an exogenous polygalacturonase is added to depolymerize the pectic acid before significant complexing with Ca^{++} can take place.

Interestingly, the formation of a Ca^{++}-pectate complex is also an important aspect in maintaining the quality of many cooked or blanched vegetables, especially those that are subsequently canned. Canned tomatoes treated with $CaCl_2$ have been observed to firm up better than raw tomatoes. This is the result of pectin methylesterase forming pectic acid and the acid moieties of the pectic acid complexing with Ca^{++}. This complex formation leads to desired firming in many vegetables in addition to tomatoes, including potatoes, beans, peppers, and cauliflower.

Another property peculiar to citrus fruits is that they contain oils that are very high in flavor and aroma compounds. These *essential oils* are quite valuable as essences and flavor ingredients. However, extraction of the oils is complicated by the fact that they form water-protein-oil emulsions that are difficult to separate. A mixture of pectinase and protease enzymes can effectively hydrolyze the pectin and protein components, making the separation of the oil from the water phase more efficient.

As we have seen, the majority of situations in which enzymes are used in the beverage industry involve the clearing of haze in the product or increasing the ease of filtration and/or extraction of a liquid. These factors influence the efficiency of production as well as the perceived quality of the final product.

Cloud loss—Precipitation of the cloudy portion of citrus fruit juice.

Essential oils—The oil fraction in plants (e.g., in citrus fruit peel) having strong flavor and aroma characteristics.

Troubleshooting

The following table lists common problems, causes that relate directly to enzymes, and changes in formulation or processing that may help to remedy the problems.

BEER		
Symptoms	**Causes**	**Changes to Make**
Insufficient starch conversion	Low α-amylase activity	Add α-amylase activity. Lower mash temperature in latter stage. Lengthen latter mashing stage. Check quality of barley malt.
	Incomplete starch gelatinization	Adjust time/temperature of cooking adjuncts and/or malt.
	Settling during mashing	Use smaller particle size of ground malt or adjunct. Lower temperature in initial stages. Add heat-stable α-amylase activity.
Difficulty filtering or slow runof	Insufficient modification of barley β-glucans and pentosans	Increase β-glucanase and/or pentosanase activity in wort. Check quality of barley malt.
Slow fermentation	Insufficient starch/protein conversion	Add exoprotease/α-amylase activity. Check that starch gelatinization was complete.
Low alcohol content	Insufficient starch conversion	Add α-amylase activity. Correct temperature and time in mashing program.
Haze in beer	Insufficient protease activity or incomplete protein hydrolysis during mashing	Increase protease activity added to wort. Increase temperature during first stage of mashing process. Lengthen first stage of mashing process. Use a protease more compatible with mashing temperature/pH. Chillproof the beer.
Poor foam stability	Excessive hydrolysis of protein	Decrease amount of added protease activity. Decrease temperature of first stage of mashing. Decrease time of first stage of mashing.
Cloudiness	Insoluble proteins or polypeptides	Increase proteolytic activity. Adjust filtration steps to remove insoluble protein.
Off flavor	Diacetyl accumulation in beer	Use fermentation organism without acetohydroxy acid synthase. Add enzyme activity that accelerates breakdown of acetolactate. Decrease stirring of wort during fermentation.
	Oxidation	Add ascorbic acid, glucose oxidase, or catalase enzyme activity. Decrease oxygen available in packaged product.
	Bitterness due to peptides	Decrease amount of proteolytic activity.

WINE		
Symptoms	Causes	Changes to Make
Grapes hard to press	Grape variety high in pectic material	Use different grape variety. Treat grapes with mixture of pectinase, xylanase, and cellulase activity.
Poor extractability/ low yield	Too much pectic material	Add pectinase activity.
Slow filtration/ too viscous	Too much β-glucan	Add β-glucanase activity.
	Too much pectic material	Add pectinase activity.
Hazy or cloudy appearance	Too much pectic material	Add pectinase activity.
	Excessive protein	Add protease activity.
Too much foam	Excessive protein	Add protease activity.
Flavor too weak	Monoterpenes are bound to β-glucan	Add β-glucanase activity.
Color instability in red wine	Tannins are bound to proteins	Add protease activity.
Browning	Oxidative degradation of phenolic compounds by polyphenol oxidase	Treat with sulfur.
Flocculation of color compounds	Oxidative degradation	Treat with sulfuric acid.
NONCITRUS FRUIT JUICE		
Symptoms	Causes	Changes to Make
Poor extraction/ low yield	Presence of pectic material	Add pectinase activity.
Slow filtration/ too viscous	Presence of pectic material	Add pectinase activity.
Haze or cloudy appearance	Presence of arabinoxylans	Add pentosanase activity.
	Starch in fruit insufficiently degraded	Add α-amylase activity during heating.
Cloudy appearance in apple juice	Presence of starch agglomerates	Add α-amylase activity.
Browning	Oxidative degradation of phenolic compounds	Heat treat to inactivate polyphenol oxidase.
Juice hard to dry	Cellulose network holding water	Add cellulase activity.

CITRUS FRUIT JUICE		
Symptoms	Causes	Changes to Make
Poor extractability/poor yield	Too much pectic material	Add pectinase activity.
Cloud loss in orange juice	Presence of pectic acid-Ca complex	Add polygalacturonase to degrade pectic acid. Heat treat to inactivate pectin esterase.
Poor extraction of essential oils	Presence of pectin/protein/oil emulsion	Add exoprotease/pectinase activity.
Pressed juice too viscous	Too much pectin	Add pectinase activity.
Poor dehydration	Presence of water-soluble pectins	Add pectinase activity.
Semigelled mass	Too much pectin	Add pectinase activity.

Reference

1. Hoseney, R. C. 1994. *Principles of Cereal Science and Technology*, 2nd ed. American Association of Cereal Chemists, St. Paul, MN.

Supplemental Reading

1. Fox, P. F., Ed. 1991. *Food Enzymology.* Elsevier Applied Science, New York.
2. Godfrey, T., and Reichelt, J. 1983, *Industrial Enzymology: The Application of Enzymes in Industry*. The Nature Press, New York.
3. Kruger, J. E., Lineback, D., and Stauffer, C. E., Eds. 1987. *Enzymes and Their Role in Cereal Technology.* American Association of Cereal Chemists, St. Paul, MN.
4. Reed, G., Ed. 1975. *Enzymes in Food Processing*. Academic Press, New York.
5. Schwimmer, S. 1981. *Source Book of Food Enzymology.* AVI Publishing Co., Westport, CT.
6. Whitaker, J. 1972. *Principles of Enzymology for the Food Sciences*. Marcel Dekker, New York.

CHAPTER 7

Other Applications

Commercial Sweetener Production

Chapter 3 discussed enzymes that convert starch to *dextrins* and simple sugars, particularly glucose and maltose. The present chapter considers the conversion of starch to sugars, primarily glucose and fructose, on an industrial level. The starch can be derived from a number of sources such as wheat, corn, rice, potatoes, and barley. By far the most common starting material is corn starch. The major classes of sweeteners produced by conversion of corn starch are corn syrups, dextrose, high-fructose corn syrup, and fructose.

The conversion of starch to sugar is generally described by referring to a unit called a "dextrose equivalent" (DE). (The terms *dextrose* and *glucose* both refer to the same sugar molecule.) A "DE" is a measure of the degree of hydrolysis of the syrup product. Starch itself has no free dextrose and so has a DE of zero. Dextrose has a DE of 100; i.e., it is 100% dextrose. All DE values are therefore between 0 and 100.

The initial (acid/heat) method of conversion of starch to sugar had a number of drawbacks: low efficiency, e.g., low rate of conversion and incomplete hydrolysis (DE of 35–45); undesirable by-products; and the need for specialized equipment to withstand the high heat under acid conditions. These negative aspects led to the search for a better conversion process.

Over the past 25–30 years, a new approach has evolved that involves the enzymatic conversion of starch to sugar. Initially, the enzymes were used in conjunction with the acid conversion. The starch slurry was gelatinized and treated with acid and heat. Then, the enzymes, usually a mixture of α-amylase and glucoamylase (amyloglucosidase) enzymes, were added to break the starch dextrins down further. It was determined that the acid pretreatment facilitated the subsequent enzyme treatment of the starch dextrins. The primary limitation with the enzymatic conversion was that, at the time, the α-amylase enzymes were still not very stable under the relatively high heat conditions used. The DE values of the sugar syrups were in the range of 60–80.

Procedures were developed that used only enzymatic starch conversion, but the rates of conversion were still not as high as desired, again because the bacterial α-amylases then in use did not have adequate heat stability. In the early 1970s, bacterial α-amylases were

---

In This Chapter:

Commercial Sweetener Production

Dairy Applications
  Cheesemaking
  Surface-Ripened Cheeses
  Enzyme-Modified Cheeses

Protein Modification

Endogenous Enzymes as Processing Indicators

Dextrins—Low molecular weight polymers of glucose (usually less than 10 glucose units in length) resulting from the enzymatic degradation of starch.

Liquefaction—The reduction in viscosity of a gelatinized starch slurry, usually accomplished using a heat-stable α-amylase of bacterial origin.

Saccharification—Further enzymatic hydrolysis of a liquefied starch slurry to produce sugars, accomplished using glucoamylase and/or pullulanase enzymes.

found that were unusually stable to heat; an example is the α-amylase from *Bacillus licheniformis*. The higher conversion rates led to an all-enzyme conversion process, resulting in sugar syrups with DE values in the range of 95–98 DE.

This all-enzymatic conversion of starch to sugars is basically a two-step process. First comes the *liquefaction* step, followed by a *saccharification* step (Fig. 7-1).

Initially, the starch is gelatinized by heating, producing a thick and viscous mixture. The temperature required to completely gelatinize the starch depends on the source. Waxy maize starch, a corn starch, can be gelatinized effectively at about 110°C (230°F). The liquefaction step is the initial thinning of this starch slurry. The subsequent addition of a heat-stable bacterial α-amylase decreases the viscosity by converting the gelatinized starch into a mixture of linear and branched dextrins along with some sugars, mostly glucose and maltose. Sugar syrups of up to about 40 DE can be achieved using bacterial α-amylases alone. But, for very high rates of conversion, a saccharification step is required. This is the step that results in a syrup composed almost entirely of sugars.

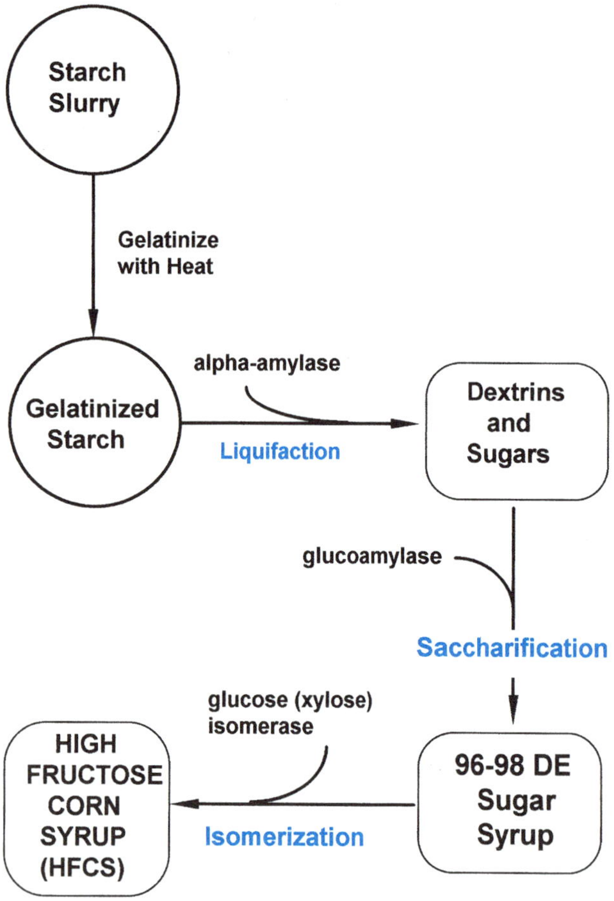

The term *saccharification* comes from the word *saccharide*, which is derived from the Latin *saccharum*, meaning *sugar*. In this step, a glucoamylase is added to the mixture of dextrins produced by the α-amylase. Depending on the glucoamylase used, a pullulanase enzyme may also be added to more efficiently hydrolyze the α,1-6 bonds at the branch points in amylopectin. The result of this saccharification step is a sugar syrup with a DE in the range of about 96–98.

The primary use for sugar syrups is as sweeteners. However, some sugars are sweeter than others. In order of decreasing sweetness, the common sugars used in the food industry are fructose, sucrose, glucose, enzyme-converted syrup, acid-converted syrup, maltose, and lactose. Because fructose is the sweetest of the common sugars, it is the most economical to use. As a result, most high-glucose sugar syrups are converted to high-fructose corn syrup (HFCS). This material is the result of another enzymatic conversion of the glucose syrup derived from the liquefaction and saccharification of corn starch.

The conversion of glucose to fructose is an example of an enzymatic reaction called *isomerization*. Isomers are chemical compounds that contain the same number and kind of

Fig. 7-1. Enzymatic conversion of starch to high-fructose corn syrup. DE = dextrose equivalent.

atoms, and have the same molecular weight, but are different in terms of their structure. The chemical formula for both glucose and fructose is $C_6H_{12}O_6$. However, their structures are different, as shown in Figure 7-2. Glucose can be isomerized into fructose through the action of an enzyme called glucose isomerase. To be more precise, this enzyme is actually a D-xylose isomerase with a secondary activity capable of using glucose as a substrate. In any case, it is able to convert some, but not all, of the glucose into fructose. The two most commonly available types of HFCS contain 42 and 55% fructose. The food industry has historically used mostly 42% HFCS, but the use of 55% HFCS is increasing.

In the past, isomerization used a batch process. This is a relatively expensive way to produce the reaction since the enzyme can be used only once with each batch. To separate the enzyme from each batch in order to use it again would prove to be prohibitively expensive. In the 1960s, a technology was developed in which enzymes were attached to a solid support while still retaining their catalytic activity. In this way, enzymatic reactions could be produced by passing a liquid stream of substrate over the *immobilized enzyme*. The reaction would proceed, and products were then carried through the reactor and collected while the enzyme remained in place. This proved to be a very important development for the starch-sugar industry because it was a much more economical method for processing high-glucose syrups into HFCS. Most HFCS production is now accomplished using continuous fixed-bed reactors in which the glucose (xylose) isomerase has been immobilized. The high-glucose syrup is passed through the reactors, resulting in a much more rapid and efficient isomerization of glucose to fructose. The rate of conversion depends on many factors, including temperature and the length of time the glucose syrup is in contact with the enzyme (i.e., the fluid flow rate). Efficient conversion requires close control and supervision. The saccharization step can be done in a similar way, using columns of immobilized glucoamylase.

Both high-glucose and high-fructose syrups find application in the food industry, primarily as sweeteners, flavor enhancers, and fermentation substrates. HFCS is used extensively in baked products, beverages, and dairy products.

Isomerization—The interconversion of different isomeric forms of the same compound (such as the enzymatic conversion of glucose to its isomer, fructose).

Immobilized enzymes—Enzymes that are chemically bound to an insoluble support material in such a way as to maintain the normal activity of the enzyme.

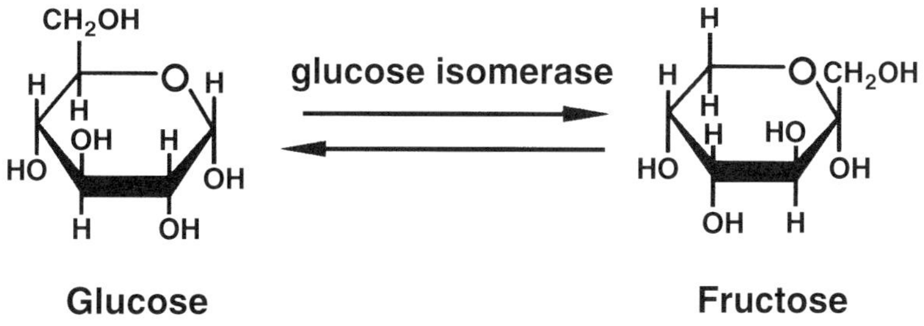

Fig. 7-2. Isomerization of glucose and fructose.

Dairy Applications

The manufacture of cheese is probably the most important application of enzymes in the dairy industry. Cheese is a product made from the curd of the milk of cows, goats, and other ruminant animals. The curd comes from the coagulation, or curdling, of the milk protein, casein. The many different kinds of cheese produced all over the world result from a number of factors including the type of milk used (cow, goat, etc.) and the methods for curdling the milk and ripening the pressed curd. The coagulation of the casein can be accomplished by adding acid to the milk or by adding an enzyme called *rennin* (also called chymosin).

Traditionally, rennin was taken from the stomachs of young ruminants. Of course, young goat stomachs are not the primary source of rennin any more. The large-scale industrial production of cheeses demand a larger, more consistent source of the enzyme. In the 1960s, the search for an enzyme similar to rennin but derived from microbial sources was begun. Several such enzymes were ultimately found and are commercially available. The commercial enzyme is sometimes called *rennet*.

CHEESEMAKING

The first step in the process of cheesemaking is the coagulating of the milk proteins. Casein is the primary protein component of ruminant milk. It is not one protein, but rather a set of proteins that associate to form small spheres of protein called "micelles." Rennin, a protease enzyme, hydrolyzes the casein protein just enough to break up the micelle and cause the protein to become insoluble and precipitate as a curd. The ability of the rennin to continue hydrolyzing the casein is restricted. It has limited specificity and does just the amount of hydrolysis needed for this process. Other proteases are also able to coagulate milk proteins, but most will continue to hydrolyze the casein, resulting in the formation of off-flavors (bitter peptides) in the finished cheese.

After the coagulation step, the curd is pressed, removing the whey (liquid fraction), then cut and cooked. The dry curds are set aside to ripen. As the cheese curd ripens, proteases, esterases, and lipases, which come from the milk or the starter culture and/or are added as exogenous enzymes, act on the protein and fat components of the curd, contributing to the taste and texture of the final product. Lipases, in particular, contribute to flavor characteristics by converting the milk fat into free fatty acids. The flavor characteristics of many Italian cheeses are the result of the use of esterases and lipases. These enzymes contribute greatly to the distinctive flavors of cheeses such as Romano, Gorgonzola, and Roquefort. These types of enzymes are also found in fungi such as *Aspergillus*, *Mucor*, and *Rhizopus*. Proteases may also be added to young cheeses to accelerate the ripening process, reducing the storage times and thus the overall cost of cheese production.

Rennin—An acid protease originally extracted from ruminant stomachs, used chiefly in cheese processing. A similar enzyme, **rennet**, derived from bacterial sources, is usually used for commercial-scale production.

SURFACE-RIPENED CHEESES

Soft or surface-ripened cheeses such as Brie and Camembert are ripened by the action of select strains of molds growing on the outer surfaces of the fresh cheese. As the mold grows, it releases proteolytic enzymes and other metabolic by-products into the cheese. The protein in the cheese then serves as a substrate for the enzyme reactions. With time, the cheese structure is broken down to a thick liquidlike consistency. At the same time, compounds are released that contribute to the flavor of the ripened cheese. If the protein breakdown is excessive, a runny consistency and an ammonia-like flavor and aroma develop.

ENZYME-MODIFIED CHEESES

Enzymes can be used to further modify cheeses, enhancing their flavor considerably and resulting in a product that can be used as a flavoring ingredient. Enzyme-modified cheeses are produced by additional enzyme modification of natural cheeses or milk solid slurries (casein). This is accomplished through supplementation with a mixture of enzymes, especially lipolytic activity and proteases. These enzymes tend to accelerate the enzymatic reactions that occur during normal cheese ripening, leading to enhanced flavor characteristics. The amount of esterase activity in the supplemental enzyme mixture is usually several times that of the lipase activity. The esterase enzymes hydrolyze the shorter-chain water-soluble fats, while the lipases liberate fatty acids from the longer-chain water-insoluble fats. As is true in traditional cheeses, addition of proteases, particularly neutral bacterial proteases, accelerates the ripening process. The use of acidic fungal proteases from *Aspergillus* sp. facilitates the development of intense flavors when the starting material consists of processed cheese waste material. The enzyme-modified material is mixed, pasteurized, and dried to produce a highly flavored food ingredient.

Protein Modification

Similar enzyme treatments on other protein material also produce highly flavored food ingredients. A widely used source of protein for this purpose is soybeans, the protein from which contains high levels of glutamic acid. Soy protein isolate may be hydrolyzed through fermentation (i.e., for soy sauce) or by addition of exogenous proteases to produce products such as hydrolyzed vegetable proteins. Most commercially available protease enzymes are effective at hydrolyzing soy protein, but the action of these enzymes does not produce the same product. Enzyme specificity plays a significant role in determining the flavor characteristics of the final hydrolyzed soy product. In a related application, the bean flavor associated with soy milk can be removed by gentle hydrolysis using neutral bacterial proteases at low temperature (about 45°C). Wheat gluten has also been used as a substrate for proteolytic hydrolysis by acidic fungal enzymes to produce a highly flavored product.

Proteins are an important contributor to the nutritional quality of many food products. However, the optimal functionality and nutritional value are not always inherent in the native protein. Proteolytic modification of native proteins is now common as a means of increasing their nutritional availability. Soy protein is an excellent candidate for enzymatic modification because, compared to other cereal proteins, particularly wheat, it has significantly higher protein content (up to 38%), contains a more complete compliment of essential amino acids (especially lysine and tryptophan), and is a readily available source of protein.

Proteolytic hydrolysis can also be used to improve the functionality of proteins. The functional properties of proteins include their ability to form gels and to retain moisture, the viscosity of the protein in solution, the dispersibility and/or solubility of the protein, their whipping characteristics, and the related ability to form a stable foam. These characteristics determine a protein's suitability for use as emulsifiers, emulsion stabilizers, moisture retention ingredients, and binders in various food products. Protease treatment of soy protein increases the whippability of the protein and the stability of the resulting protein foam, thus improving its functional properties in food. Whey proteins have a much increased degree of solubility as a result of protease treatment, allowing them to be used in a variety of foods, especially as a skim milk replacement for lactose-intolerant individuals. Increasing the solubility also permits these proteins to be used as protein supplements in carbonated and other acidic beverages. Hydrolyzed soy protein has been formulated as an alternate milk product. Soy protein as well as milk protein fractions have been made into many types of beverages used as nutritional supplements.

Improving the food functionality of high-quality proteins such as those from soybeans or milk increases their use in a variety of food products. Of course, these modifications also affect the nutritional quality of the proteins. Consequently, a greater variety of proteins, with a different, more complete complement of amino acids is made available to more people.

Endogenous Enzymes as Processing Indicators

Enzymes can also serve as indicators of the presence of certain compounds and as a means of determining the desired endpoint in a processing scheme. Although an enzyme is usually used to modify a substrate for a particular purpose, if the reaction takes place and can be monitored, it also serves as an indicator that the substrate was, in fact, present. Thus, enzymes are used as indicators for the presence of such food constituents as alcohol, lactose, lecithin, various sugars, starch, and cholesterol.

Pasteurization and blanching processes are designed to inactivate certain enzymes and/or microorganisms by heat denaturation. Inactivation of peroxidase (EC 1.11.1.7) has been found to be a good indicator of the extent of blanching (of vegetables) that inactivates endogenous deteriorative enzymes without significantly damaging the food texture. Complete inactivation of all peroxidase activity is undesirable—the vegetables are adversely affected. The proper treatment is achieved when about 5–10% of the peroxidase activity remains. Of course, the ability to accurately determine this point is dependent on the sensitivity of the method for assaying the peroxidase activity. The accepted method involves the use of guiacol and colorimetric determination of the enzymatic product, tetraguiacol (see Appendix D, peroxidase assay no. 1). This protocol has been judged to be adequate for blanching of vegetables such as peas, corn, beans, and brussels sprouts.

Peroxidase activity is also used as a diagnostic means of determining the correct extent of heating oats to destroy lipid-degrading enzymes that can lead to rancidity in the oat product. Again, the inactivation of peroxidase serves as the indicator of thermal inactivation of the lipolytic enzymes.

In a related application, alkaline phosphatase (EC 3.1.3.1) is used as the measure of adequate pasteurization of milk products. The goal of pasteurization is to kill potentially harmful microorganisms in the milk. It is a coincidence that the phosphatase enzyme endogenous to milk is heat-inactivated in the temperature range necessary to kill the offending organisms. Thus, the enzyme is used as a convenient measure of pasteurization adequacy. This test is a simple colorimetric assay using phenol phosphate as the substrate (see Appendix D, assay for alkaline phosphatase).

Although soybeans are an excellent source of protein, to derive the most nutritional benefit from them, particularly when they are used as feed, the beans must be heated to alter the flavor and inactivate the antinutritional factors (trypsin inhibitors) they contain. Under- or overcooking produces undesirable effects, so the correct degree of cooking is determined by the inactivation of urease (EC 3.5.1.5).

Supplemental Reading

1. Fox, P. F., Ed. 1991. *Food Enzymology*. Elsevier Applied Science, London.
2. Hettiarachchy, N. S., and Ziegler, G. R., Eds. 1994. *Protein Functionality in Food Systems*. Marcel Dekker, New York.
3. Messing, R. A., Ed. 1975. *Immobilized Enzymes for Industrial Reactors*. Academic Press, New York.
4. Schwimmer, S. 1981. *Source Book of Food Enzymology*. AVI Publishing Co., Inc., Westport, CT.
5. Weetal, H. H., and Suzuki, S. 1974. *Immobilized Enzyme Technology: Research and Applications*. Plenum Press, New York.
6. Whitaker, J. 1977. *Food Proteins: Improvement Through Chemical and Enzymatic Modification*. American Chemical Society, Washington, DC.

CHAPTER 8

Choosing Enzymes for Specific Applications

An Approach to Solving Problems

Knowledge of enzymes and how to use them in food processing has become significantly more sophisticated in the last decade or so. The more the conditions of production are understood, the better the solutions that can be conceived for solving specific production issues. While the range of problems associated with different processes and finished products is indeed very wide, there are some basic approaches to determining the fundamental issues and addressing them in a logical, progressive manner. Regardless of the particular industrial application, asking the right questions should help you to proceed to a solution.

This chapter looks at some examples to illustrate an approach useful to a person involved in developing a new product or improving an existing one.

WHAT IS THE PROBLEM?

You might be developing a totally new beverage, a baked product with a fruit filling, or a new flavor base from a canola seed—or perhaps something less intimidating such as an improvement to the texture of a cracker, increasing the extractable yield of oils, or clarifying a meat broth. Whatever your particular food area, you should try to describe the objective of your project as precisely and completely as possible. You may need to consult with others involved in the production of the product in order to understand as completely as possible all the factors involved.

> **In This Chapter:**
>
> An Approach to Solving Problems
>
> New Opportunities for the Future

> **Questions to Define Use of an Enzyme**
> 1. What is the problem? Define it as thoroughly as possible.
> 2. What is the likely source of the problem?
> 3. What are the major components of the food system?
> 4. Is use of an enzyme the appropriate solution?
> 5. Which components of the food system need to modified?
> 6. What is the pH of the food system to be modified?
> 7. What are the temperature constraints of the food system?
> 8. What are the time constraints in the process?

WHAT IS THE LIKELY SOURCE OF THE PROBLEM ?

Once you have defined the project and/or the desired goal, you must determine whether the desired goal is the result of the formulation used, the processing conditions, or an interrelationship between the two. As an example, suppose that your product is formulated to have an acidic pH to provide a certain taste profile. You have included a protein to add body to the drink, but the protein becomes insoluble at the pH of the final product. The protein precipitates, producing cloudiness in the beverage or clogging the delivery lines. In this case, the problem can be identified as a formulation issue. In another instance, dough sheets may stick to the conveyor bands, making separation of the scrap from the cut piece difficult. This could be caused by improper band material or by a formulation that makes the dough unusually tacky. This implies either a process issue or a formulation problem. You would need to determine which approach would best serve the final product you wish to produce. In any case, the problem will be solved more efficiently if you can identify whether you have a process problem, a formulation problem, or possibly a combination of both.

WHAT ARE THE MAJOR COMPONENTS OF THE FOOD SYSTEM?

Most foods have a limited number of components that have major effects in the production of a product. If the overall formulation itself does not seem to be an obvious problem, what about the ingredients themselves? How much water is in the formulation and what is its temperature when it is added? Is the pH of a liquid product changing as the result of addition of another ingredient? Is the flour appropriate for the product? What is the protein content? Is the order of addition of the ingredients important? Are any fermentation organisms viable and active? Try not to make the project more complicated than it is. Identify and concentrate on the primary components first and determine their principal function(s) in the product. What are they likely to contribute to the processing conditions and to the final product characteristics?

IS USE OF AN ENZYME THE APPROPRIATE SOLUTION?

Now you must evaluate whether enzymes represent a viable approach to addressing the issues at hand. This is a very important question because, in many cases, it may be best not to use enzymes in a process. Enzymes increase the cost of production and introduce another functional ingredient that should then be subject to quality assurance/quality control procedures. If it is possible to solve the problem without using enzymes, don't use them. For example, the shape and texture of a baked product are primarily functions of the flour protein. If the protein content of the flour is too high for a particular product, a protease enzyme could be added to mellow the dough, but that may not be the best approach. If a flour with a lower protein content is available and will work satisfactorily, that is a better, cheaper, more consistent solution. Processing issues such as the mixing time or the quantity and temperature of the water may be adjusted to help alleviate the problem.

In other situations, processing conditions may be inactivating an endogenous enzyme that would otherwise contribute to the desired characteristics. For example, the current processing temperature may be too high for a new enzyme preparation, requiring some modification of the current process. Thus, the answer could lie in the individual ingredients or in particular processing conditions rather than in the application of an exogenous enzyme.

WHICH COMPONENTS OF THE FORMULATION NEED TO BE MODIFIED?

Assuming that a decision has been made that at least one of the components needs to be enzymatically modified to alleviate the problem, what type of enzyme do you use? The choice depends on the component to be altered, the pH of the formulation, the temperature during processing, and the length of time available for exposure to enzymatic activity.

If you decide that not just one component must be enzymatically modified, another decision presents itself. Should you use separate enzyme preparations or one preparation with all the enzymes mixed together? Either alternative may provide a suitable solution. The use of separate enzymes provides the best control, i.e., the most consistent enzyme activity over time. However, it may be more expensive than a single mixture of enzymes.

It is very important to remember that enzyme suppliers do not standardize their preparations on all activities present in the mixture. They may acknowledge that a certain enzyme preparation has the types of enzymes you need but standardize on only one (usually the major activity present). This could lead to problems since the level of one of the enzymes you need may vary over time and change the processing/product characteristics. Ideally, your specifications should require that the preparation contain only the activity that your process requires or that the activities of all enzymes contained in the preparation be measured and reported. That way, you can be reasonably sure that the next shipment of enzymes you receive will function the same as the last one.

The proper level of enzyme to be used in your formulation is best determined at your facility, under the same conditions that you anticipate using for your product. Changes in conditions may require changes in the enzyme level used. Start with the recommended use level provided by your supplier, but do not depend on it as final. The level of enzyme used will be a function of the degree of modification required for the food component(s) and the length of time available for the modification. Try to find a point in the process that allows the most time for enzymatic modification. This will permit the minimum amount of enzyme to be used, keeping your costs down and minimizing subsequent control issues. You should also try to establish the relationship between the amount of enzyme used, at a particular activity level, and the size of the product batch. This makes scaling easier if you alter the batch size or other production variables. Also, keep careful records of the amounts of enzyme used in the experimental trials.

Remember that your best insurance for consistent enzyme products is to have your own internal quality control/assurance program that includes the methodology for independent verification of enzyme activities. Your method may be different from that used by the enzyme supplier and should reflect the substrates and conditions actually used at your facility. The results of the supplier should correlate with your own test results. The ability to independently evaluate these functional ingredients provides you with effective accept/reject criteria for the enzyme shipment as well as a system that can contribute to the consistent production of your products. It also allows you to have several sources of supply of enzymes without wondering whether the different enzyme preparations can be used interchangeably.

WHAT IS THE pH OF THE FOOD SYSTEM TO BE MODIFIED?

The choice of enzyme may depend on where and when in the process it is applied. In some processes, the pH does not change significantly from beginning to end; on the other hand, a procedure such as soda cracker production involves a sponge and dough bacterial fermentation in which the pH during fermentation of the sponge becomes increasingly acidic but later in the process becomes basic. You must determine where in the process the enzyme is to be added and use the enzyme best suited to the pH at the time of use.

Operating at a pH significantly out of the optimal range of your desired activity results in a poor enzymatic response. This may encourage you to add more enzyme to speed the rate of modification, which will have several undesirable consequences. The more enzyme you add, the more difficult it is to control the activity and the higher the cost. Remember, once you put the enzyme in, you can't take it out. Also, under normal circumstances, it is very difficult to quickly change the environment of the product mass as a means of controlling the enzyme activity. The more enzyme in the mixture, the more sensitive it is to small changes in production conditions. For example, if the temperature is a bit hotter the next time, significantly more enzymatic modification will result and the resulting product will likely be different. If something happens further along the production line to slow or halt additional processing, you will have problems because the enzymes continue to work regardless of the status of your production line.

WHAT ARE THE TEMPERATURE CONSTRAINTS OF THE FOOD SYSTEM?

Another parameter critical to the success of enzymatic modification is the temperature of the food system at the point you wish the enzyme to work. Most enzymes applicable to foods work in a temperature range of about 40–60°C. You should be aware of any temperature changes in the processing conditions and plan to use the appropriate enzyme.

The reasons for paying attention to the temperature issue are similar to those described for pH. The main functional activity as well as any side activities will respond to changes in temperature. It is best to accomplish the enzymatic modification during a step that is at a constant temperature and allows adequate time for the extent of modification required.

Another, related, issue that must be considered is whether the enzyme can be left in the product in an active condition. Enzymes are commonly allowed to perform their task and are then inactivated to prevent any further alteration in the product. For example, if the amylase used in bread production is from a bacterial rather than a fungal source, its thermal stability may allow it to remain active throughout baking, resulting in extended modification not desired or anticipated. On the other hand, if you wish to minimize the concentration of oxygen to prevent oxidative changes in your product, you may want to retain enzymatic activity (glucose oxidase for example) through as much of the processing of the product as possible, as well as during storage.

WHAT ARE THE TIME CONSTRAINTS IN THE PROCESS?

After ensuring that the enzyme is suited to the pH and temperature of your food system, you must make sure that it has sufficient time to modify the targeted food component(s). It is better to have plenty of time for a small amount of enzyme to act than to add large amounts of enzyme to achieve sufficient modification in a short time. The more enzyme you add to your formulation, the more critical the time factor becomes. With large amounts of enzyme activity at work, timing becomes very important. If something goes wrong somewhere down the production line, causing delays, the entire batch could possibly be lost due to overmodification.

The production process for baked products often involves a rest period or what is sometimes referred to as "lay time" or "floor time." This is a good opportunity for enzyme use because in the dough at rest the components are not free to migrate throughout the vessel holding the dough. In general, most of the enzymatic modification that is going to take place will do so within the first 45–60 min. Additional modification then proceeds at a slower, decreasing rate as all the substrate material available to the enzyme becomes modified. Unless the dough is disturbed further, by remixing, for example, the enzyme cannot migrate far enough to encounter new substrate. If the proper amount of enzyme was added to the formulation, that is, the minimum amount required to achieve the intended result in the prescribed time period, overmodification of the dough should not be a major concern unless there is a significant delay in processing and/or the dough is disturbed in such a way as to redistribute substrate and enzyme.

Systems in which the target of enzyme activity is in liquid solution, or even in a batter, are more susceptible to extended enzyme activity over time. You must assume in these situations that the enzyme is capable of complete hydrolysis of the substrate given enough time. Under these conditions, timing of the process must be well controlled if you want to achieve less than complete alteration in the formula component(s).

New Opportunities for the Future

This book has addressed the currently developed enzyme technology used in processing foods and beverages. While this represents a significant amount of research and development over many years, considerable potential remains for continued development of enzyme technologies applicable to the food industry.

For example, utilizing the specificity of enzymes has the potential to provide new opportunities. A major concern when using proteases (as well as other enzymes) is that use of too high a level of the enzyme can have a potentially negative effect on both the process and the product. We could benefit from enzymes that give just the amount of modification required and no more. This is not as far-fetched as it may sound. Cheese production is much less problematic than many other processes because chymosin has a limited specificity. In normal use, the enzyme will not overmodify the milk protein. Other food processes could also benefit from enzymes having limited specificity, resulting in only the required degree of modification and thus providing a built-in control mechanism.

A similar concept concerns the use of disulfide isomerases or oxidases in the baking industry. At least part of the viscoelastic nature of gluten is attributed to disulfide bond formation. More work could be done to investigate enzymes affecting the disulfide-sulfhydryl interchange in wheat doughs as a means of controllably modifying gluten texture. There are thousands of microorganisms available to be investigated, and looking for such enzyme characteristics is a worthwhile pursuit. It may be better to strive to develop sources for more specific enzyme activities rather than to create more broad-specificity mixtures.

The increasing use of high-speed processing, as exemplified by extruders and continuous mixers, suggests that the time allowed for enzymatic modification of a given product may become limited. There will then be a need for specific, fast-acting enzymes that can operate under pressure and in a relatively low-moisture environment. Currently available enzymes may not be suitable for such production conditions.

The use of immobilized enzymes has revolutionized the starch conversion industry, and more applications for them could be devised for other food processes. For example, instead of adding enzymes to a complete formulation when only a particular fraction of the mix requires modification, pretreatment of the target ingredient may be a more efficient approach. For example, a flour-water suspension might be pretreated before use in a formulation. An extension of this idea involves the use of enzymes to produce desired components (such a flavor precursors) in situ as opposed to adding those components as labeled ingredients.

Discussions of enzyme use generally assume that a catabolic process is desired. However, enzymes are also capable of anabolic action. In that regard, new opportunities exist for enzymes such as lipases and proteases. Using the process of interesterification, lipases can be used to produce new "designer fats," which could be synthesized to fit the precise requirements of a particular application. Lipases have also shown anabolic activity in nonaqueous media. As for proteases, the plastein reaction, which involves the synthesis of proteins from constituent amino acids and peptides using proteases, has been known for some time but has not yet been utilized to a significant extent. It also represents a potentially fruitful area—in this case the production of "designer proteins." The bases for such work have been established both through computer modeling of new proteins (currently used in the pharmaceutical industry) and through research on the enzymatic synthesis of peptides and small proteins.

The development of such capabilities will require concerted effort and cooperation between food producers and enzyme manufacturers. The food producers must make clear to the enzyme suppliers what is required so that enzyme manufacturers can best apply their expertise to finding the specific enzyme required.

APPENDIX A.
Amino Acid Side Chains[a]

Non-Polar Side Chains

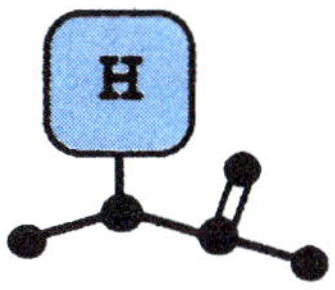

Glycine — H

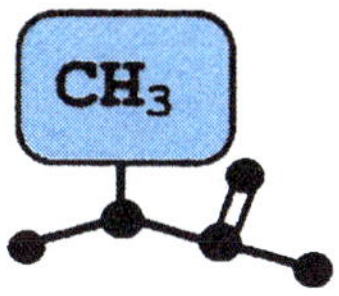

Alanine — CH_3

Valine — $(CH_3)_2CH$

Proline

Leucine — $(CH_3)_2CH-CH_2$

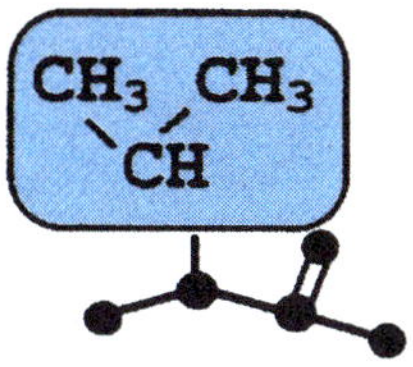

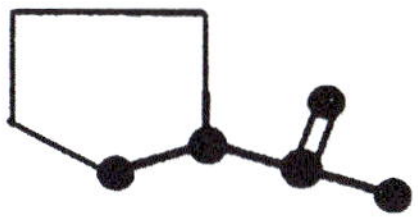

Isoleucine — $CH_3-CH_2-CH(CH_3)$

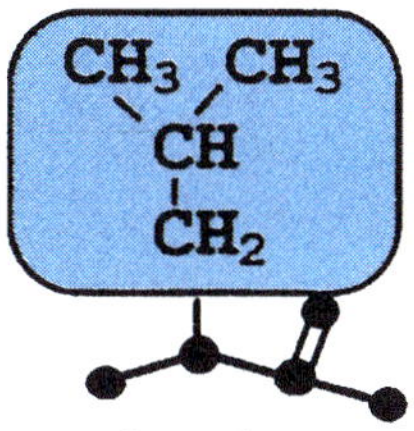

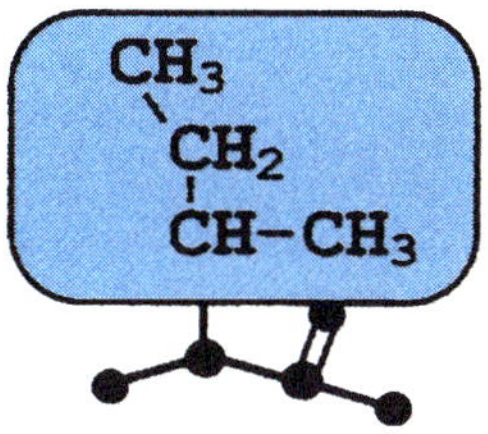

Phenylalanine — $C_6H_5-CH_2$

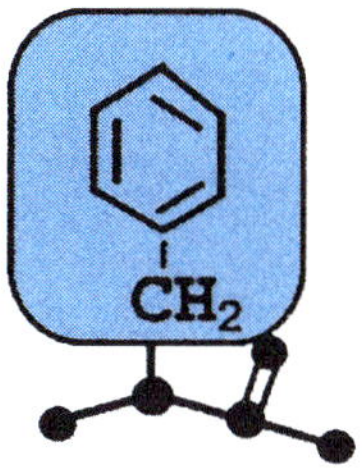

Methionine — $CH_3-S-CH_2-CH_2$

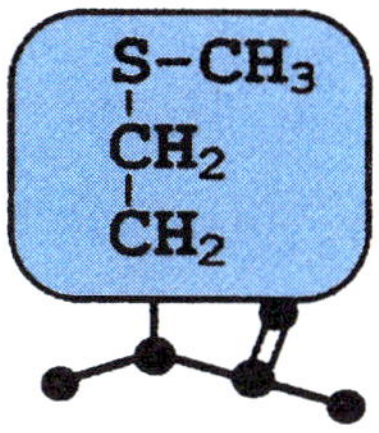

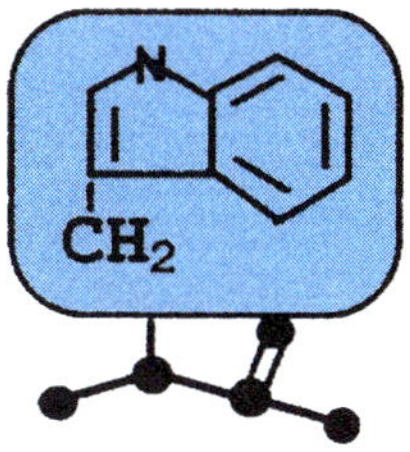

Tryptophan — indole-CH_2

Acidic Side Chains

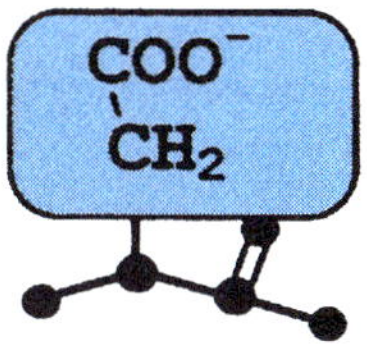

Aspartic Acid — COO^--CH_2

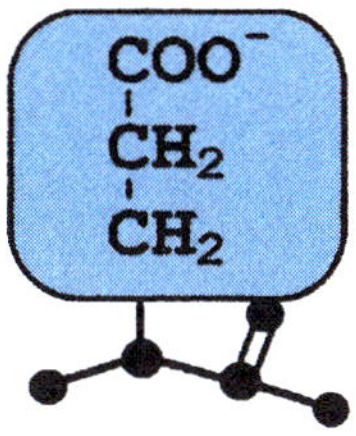

Glutamic Acid — $COO^--CH_2-CH_2$

[a] Charges indicated at pH 7.0.

Amino Acid Side Chains, continued

Neutral Side Chains

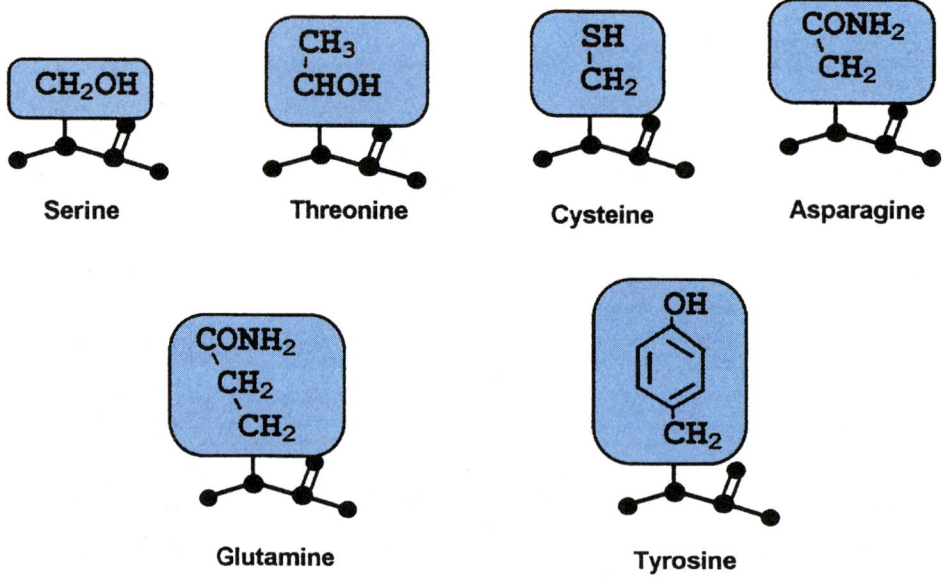

Basic Side Chains

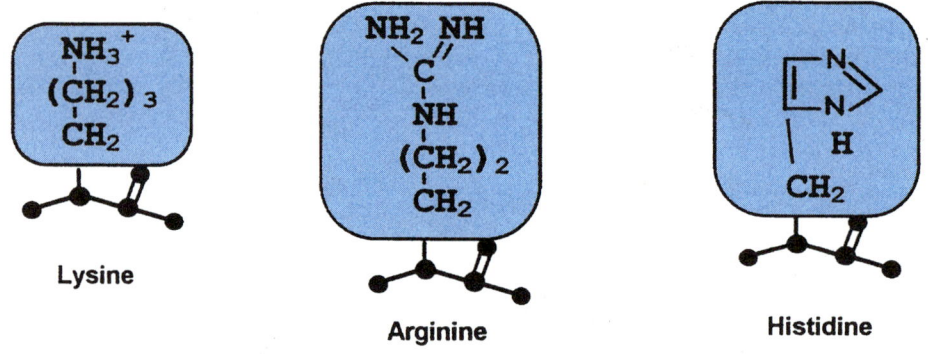

APPENDIX B.
Description of Selected Enzymes

Enzyme Class	Common Name	EC Number	Substrate	pH Range	Optimum Temperature (°C)
Oxidoreductase	Glucose oxidase	1.1.3.4	D-Glucose	5–7	30
	Catalase	1.11.1.6	Hydrogen peroxide	3–9	25
	Peroxidase	1.11.1.7	Many hydrogen-donating compounds	6–8	25
	Lipoxygenase	1.13.11.12	Polyunsaturated fatty acids	7–8	25
	Phenol oxidases	1.10.3.1	Catechol + oxygen		
		1.10.3.2	p-Diphenol + oxygen	5–8	25
		1.14.18.1	Tyrosine + oxygen		
Hydrolases	Lipase	3.1.1.3	Free fatty acids	5.5–8.5	30
	Pectinesterase	3.1.1.11	Methyl ester of pectin	4.5–7.5	30
	Phytase	3.1.3.8	Phytic acid	5–7	55
	α-Amylase	3.2.1.1	Starch	5–7	60
	β-Amylase	3.2.1.2	Starch	5–7	60
	Glucoamylase	3.2.1.3	Oligosaccharides with α-1,4 and -1,6 links	4–5	55
	Cellulase	3.2.1.4	Cellulose	4.5–6.5	45
	β-Glucanase	3.2.1.6	1,3- (1,4-) β-Linked glucan	4–6	45
	Endopolygalacturonase	3.2.1.15	Pectic acid	3.5–4.5	35
	Exopolygalacturonase	3.2.1.67	Pectic acid	3.5–5.0	35
	Invertase	3.2.1.26	Sucrose	4–5	55
	Xylanase	3.2.1.32	Xylan	5.5–6.5	40
	Pullulanase	3.2.1.41	α-1,6-linked oligosaccharides	4.5–5.5	25
	Leucine aminopeptidase	3.4.1.1	N-terminal amino acid in proteins	8–9	25
	Carboxypeptidase	3.4.2.1	C-terminal amino acid in proteins	7–8	25
	Microbial serine protease	3.4.21.14	Protein	7–9	35
	Papain	3.4.22.1	Protein	5–7	35
	Ficin	3.4.22.3	Protein	4.5–8.5	35
	Bromelain	3.4.22.5	Protein	6–8	35
	Rennin (chymosin)	3.4.23.4	Protein	3.5–4.5	35
	Microbial carboxyl proteinase	3.4.23.6	Protein	4.5–5.5	35
	Microbial metalloprotease	3.4.24.4	Protein	6–8	35
Lyases	Pectate lyase	4.2.2.2	Pectins	7.5–9	35
	Pectin lyase	4.2.2.10	Pectins	5–6	35
Isomerases	Glucose (xylose) isomerase	5.3.1.5	Glucose (xylose)	7.5–8.5	50

APPENDIX C.
Major Biological Sources of Enzymes

Source	Principal Activity of Enzyme	Regulatory Status of Enzyme[a] United States	Europe	Primary Use Code[b]
Bacteria				
Actinoplanes missouriensis	Xylose (glucose) isomerase	GRAS[c] (184.1372)	RAS[d]	Bv, Cr, St, Su
Arthrobacter globiformis	Xylose (glucose) isomerase	NA[e]	RAS	Bv, Cr, St, Su
Bacillus acidopullulyticus	Pullulanase	NA	RAS	Bk, Bv, Cf, St, Su
B. cereus	Milk clotting enzymes	CFR 173.150	NA	Bv, Ch, Bk
B. coagulans	Protease	NA	RAS	Bv, Cr, Mt
	Glucose (xylose) isomerase	GRAS (184.1372)	RAS	Cf, Cr, St, Su, Vg
B. lichenformis	α-Amylase	GRAS (184.1027)	RAS	Bk, Cf, St, Su, Vg
	Protease	GRAS (184.1027)	RAS	Mt, Bv
B. megaterium	α/β-Amylase	NA	RAS	Bv, Cr, St,
B. stearothermophilus	α-Amylase	GRAS (184.1012)	RAS	Bk, Bv, Cf, St, Su, Vg
B. subtilis	α/β-Amylase	GRAS	RAS	Bk, Bv, Ch, Cf, Ft, St
	Protease	GRAS	RAS	Bk, Bv
	Hemicellulase	GRAS	RAS	Cf
Klebsiella planticola	Pullulanase	NA	RAS	Bk, Bv, Cf, St, Su
Streptomyces olivaceus	Glucose (xylose) isomerase	GRAS (184.1372)		Bv, St, Su, Vg
S. olivochromogenes	Glucose (xylose) isomerase	GRAS (184.1372)	RAS	Bv, St, Su, Vg
S. rubiginosus	Glucose (xylose) isomerase	GRAS (184.1372)	RAS	Bv, St, Su, Vg
Fungi				
Aspergillus melleus	Protease	NA	RAS	Ch
A. niger	α-Amylase	GRAS	RAS	Bk, Bv, Cr, St, Su, Vg
	Endoarabinase	GRAS	RAS	Bv
	Glucoamylase	GRAS	RAS	Bv, Cf, St, Su, Vg
	Hemicellulase	GRAS	RAS	Cf, Sf
	Pectinase	GRAS	RAS	Bv, Vg,
	Xylanase	GRAS	RAS	Bv, Cf, Cr, St
	Cellulase	GRAS	RAS	Bv, Sf, Vg
	Catalase	GRAS	RAS	Bv, Ch, D
	Phytase	GRAS	RAS	Bk, Cr, Vg
	Protease	GRAS	RAS	Bk, Bv, Ch, Cr, Mt, Vg
	Lipase	GRAS	RAS	Ch, Ft
	Glucose oxidase	GRAS	RAS	Bv, D

(continued)

Major Biological Sources of Enzymes, continued

Source	Principal Activity of Enzyme	Regulatory Status of Enzyme[a]		Primary Use Code[b]
		United States	Europe	
A. oryzae	α-Amylase	GRAS	RAS	Bk, Bv, Cr, St, Su, Vg
	Cellulase	GRAS	RAS	Bv, Vg
	Glucoamylase	GRAS	RAS	Bk, Bv, Cf, Ft, Su
	Hemicellulase	GRAS	RAS	Cf
	Lipase	GRAS	RAS	Ch, Ft
	Pectinase	GRAS	RAS	Bv, Vg
	Protease	GRAS	RAS	Bk, Bv, Ch, Mt, Vg
Disporotrichum dimorphosporum	Cellulase	NA	RAS	Bv, Vg
	Hemicellulase	NA	RAS	Cf
	Glucanase	NA	RAS	Bv, Cf, Cr
	Xylanase	NA	RAS	Bv, Cf, Cr
Endothia parasitica	Milk-clotting enzymes	CFR 173.150	RAS	Ch
Humicola insolens	Cellulase	NA	RAS	Bv
	Hemicellulase	NA	RAS	Bk, Cf
	Pentosanase	NA	RAS	Bk
	Xylanase	NA	RAS	Bk
Mortierella vinaceae	α-Galactosidase	CFR 173.145	RAS	Su
Mucor javanicus	Lipase	NA	RAS	Ch, Ft
M. meihei	Lipase-esterase	CFR 173.140	RAS	Ch, Ft
	Milk-clotting enzymes	CFR 173.150	RAS	Ch
M. pusillus	Milk-clotting enzymes	CFR 173.152	RAS	Ch
	Lipase	NA	RAS	Ch, Ft
Penicillium emersonii	Cellulase	NA	RAS	Bv
	Glucanase	NA	RAS	Bv
	Xylanase	NA	RAS	Bv
P. funiculosum	Cellulase	NA	RAS	Bk, Bv, Ft, Vg
	Cellobiase	NA	RAS	Bk, Bv, Ft, Vg
	Glucosidase	NA	RAS	Bv
	Dextranase	NA	RAS	Su
	Glucanase	NA	RAS	Bv
	Glucoamylase	NA	RAS	Bk, Bv, Cf, Su, Vg
	Pectinase	NA	RAS	Bv, Vg
	Xylanase	NA	RAS	Bv
P. lilacinum	Dextranase	NA	RAS	Su
P. simplicissimum	Pectinase	NA	RAS	Bv, Vg

(continued)

Major Biological Sources of Enzymes, continued

Source	Principal Activity of Enzyme	Regulatory Status of Enzyme[a] United States	Regulatory Status of Enzyme[a] Europe	Primary Use Code[b]
Rhizopus niveus	Glucoamylase	CFR 173.110	NA	Bk, Bv, Cf, Su, Vg
R. oryzae	Carbohydrases	CFR 173.130	NA	Bk, Bv, Cf, St, Vg
Trichoderma harzianum	Cellulase	NA	RAS	Bv, St, Vg
	Glucanase	NA	RAS	Bv
	Glucosidase	NA	RAS	Bv
	Hemicellulase	NA	RAS	Bv, St, Vg
	Xylanase	NA	RAS	Bv, St, Vg
T. longbranchiatum	Cellulase	GRAS	RAS	Vg
	Glucanase	GRAS	RAS	Bv
	Glucoamylase	GRAS	RAS	Bk, Bv, Cf, Su, Vg
	Hemicellulase	GRAS	RAS	Cf
	Pectinase	GRAS	RAS	Bv, Vg
	Xylanase	GRAS	RAS	Bk
Yeasts				
Kluyveromyces marxianus	Chymosin (rennin)	NA	RAS	Ch
	Invertase	GRAS	RAS	Cf
K. lactis	Lactase	GRAS (184.1388)	RAS	Ch, D
Saccharomyces cerevisiae	Invertase	GRAS	RAS	Cf
Plants				
Barley, malt	α-Amylase	GRAS	RAS	Bk, Su, St
	β-Amylase	GRAS	RAS	Bk, St
	Protease	GRAS	RAS	Bk
Pineapple	Bromelain (protease)	GRAS	RAS	Bk, Bv, Mt
Figs	Ficin (protease)	GRAS	RAS	Bv, Mt, Vg
Papaya	Papain (protease)	GRAS	RAS	Bk, Bv, Cr, Vg
Animal				
Bovine, liver	Catalase	GRAS	RAS	Bv, D
Ruminants, fourth stomach	Rennin (protease)	GRAS	RAS	Ch
Porcine/bovine, stomach	Pepsin (protease)	GRAS	RAS	Bv, Ch, Cr

[a] Relevant paragraphs in the *Code of Federal Regulations* (CFR) are indicated.

[b] Application codes: Bk = baking, Bv = beverages (beer, wine, juice, soft drinks), Cf = confections (cocoa, chocolate, etc.), Ch = cheese, Cr = cereal, D = dairy (milk, eggs), Ft = fats/oils, Mt = meats/fish, Sf = seafood, St = starch, Su = sugars, Vg = fruits/vegetables.

[c] Generally Recognized As Safe (USA). Conditions for use of GRAS materials are prescribed in the referent regulations and are predicated on the use of nonpathogenic and nontoxicogenic strains of the respective organisms and on the use of current good manufacturing practice (184.1(6)).

[d] Recognized As Safe (Europe). The microorganisms listed meet one or both of the following criteria: 1) RAS based on documented history of use in food or food additives production and/or common presence in food; 2) RAS based on scientific evidence from toxicological testing, the extent of which takes into account the intended use in food processing.

[e] Not approved.

APPENDIX D.
Specific Enzyme Assay Techniques

Glucose Oxidase (EC 1.1.3.4)
Reaction: Oxidization of glucose to gluconic acid and hydrogen

1. Reference: Ciucu, A., and Patroescu, C. 1984. Analytical Letters 17:1417
   Method: Spectrophotometric
   Description: The hydrogen produced in the reaction reduces o-benzoquinone dihydroxybenzene (catechol), which absorbs at 290 nm.
2. Reference: Brocklehurst, K. 1992. Page 201 in: *Enzyme Assays: A Practical Approach*. R. Eisenthal and M. J. Danson, editors. Oxford University Press
   Method: pH-stat
   Description: Liberated hydrogen (proton) tends to drop the pH of the solution. The pH drop is sensed and recorded by the pH-stat.
3. Reference: Clark, J. B. 1992. Page 187 in: *Enzyme Assays: A Practical Approach*. R. Eisenthal and M. J. Danson, editors. Oxford University Press
   Method: Specific-ion electrode (oxygen)
   Description: The oxygen electrode senses removal of oxygen from solution as glucose is oxidized.

Catalase (EC 1.11.1.6)
Reaction: Breaking down of hydrogen peroxide to water and molecular oxygen

1. Reference: Sinha, A. K. 1972. Analytical Biochemistry 47:389
   Method: Spectrophotometric (colorimetric)
   Description: Hydrogen peroxide reduces potassium dichromate in acetic acid to chromic acetate, which absorbs at 570 nm.
2. Reference: Kroll, R. G., Frears, E. R., and Bayliss, A. 1989. Journal of Applied Bacteriology 66(3):209
   Method: Specific-ion electrode (oxygen)
   Description: The oxygen electrode senses the liberation of oxygen from H_2O_2.

Peroxidase (EC 1.11.1.7)
Reaction: Catalysis of the H_2O_2-mediated oxidation of a donor molecule with consequent decomposition of H_2O_2

1. Reference: Varoquaux, P., Clochard, A., Sarris, J., Avisse, C., and Morfeaux, J. N. 1975. Lebensmittel Wissenshaft und Technologie 8:60 (in French)
   Method: Spectrophotometric (colorimetric)
   Description: Guiacol is oxidized to a highly colored compound, tetraguiacol, which is measured by absorbance determination.
2. Reference: Ngo, T. T., and Lenhoff, H. M. 1980. Analytical Biochemistry 105:389
   Method: Spectrophotometric (colorimetric)
   Description: Oxidation of 3-methyl-2-benzothiazolinone • HCl • H_2O (MBTH) and 3-dimethylaminobenzoic acid (DMAB) by peroxidase in the presence of H_2O_2 results in the formation of an indamine dye, which absorbs at 590 nm.
3. Reference: Iwai, H. F., Ishihara, F., and Akihama, S. 1983. Chemical Pharmacology Bulletin 31:3579
   Method: Fluorometric

Description: Reactant molecules o-dianisidine and homovanillic acid form a highly fluorescent molecule upon oxidation by the enzyme. The reaction is followed at 425 nm.
4. Reference: ALTECA, Ltd. (Manhattan, KS), product literature
Method: spectrophotometric (colorimetric)
Description: This kit utilizes guiacol as the substrate. Details of the reaction are proprietary.

Alkaline Phosphatase (EC 3.1.3.1)

Reaction: Hydrolysis of an orthophosphoric ester substrate (phenyl phosphate) to the alcohol (phenol) and orthophosphate

1. Reference: Scheimann, D. A., and Brodsky, M. H. 1976. Journal of Milk and Food Technology 39:191
   Method: Spectrophotometric (colorimetric)
   Description: Phenyl phosphate is hydrolyzed to phenol and orthophosphate. The phenol is reacted with 2,6-dibromoquinone-4-chloridate (BCQ), resulting in a colored compound that can be measured by following absorbance at 610 nm.

Lipoxygenase (EC 1.13.11.12)

Reaction: Oxygenation of polyunsaturated fatty acids

1. Reference: Gibian, M. J., and Galaway, R. A. 1976. Biochemistry 15:4209
   Method: Spectrophotometric
   Description: Lipoxygenase reacts with linoleic acid, producing the fatty acid hydroperoxide, which absorbs at 235 nm.
2. Reference: Toyosaki, T. 1992. Journal of the AOAC International 75(6):1124
   Method: Spectrophotometric (colorimetric)
   Description: Enzyme activity is measured as the time required to bleach methylene blue. The reaction is followed by determining the decrease in absorbance at 660 nm.

Lipase (EC 3.1.1.3)

Reaction: Hydrolysis of triglycerides, forming free fatty acids

1. Reference: Mosmuller, E. W. J., van Heemst, J. D. H., van Delden, C. J., Franssen, M. C. R., and Engbersen, J. F. J. 1992. Biocatalysis 5(4):279
   Method: Spectrophotometric
   Description: The enzyme reacts with 2,4-dinitrophenol butyrate, forming the 2,4-dinitrophenlate anion, which absorbs at 360 nm.
2. Reference: Versaw, W. K., Cuppett, S. L., Winters, D. D., and Williams, L. E. 1989. Journal of Food Science 54(6):1157
   Method: Spectrophotometric (colorimetric)
   Description: The method uses a color reaction between Fast Blue BB and β-napthol, which is cleaved by the enzyme from β-napthol caprylate.
3. Reference: Brocklehurst, K. 1992. Page 205 in: *Enzyme Assays: A Practical Approach*. R. Eisenthal and M. J. Danson, editors. Oxford University Press.
   Method: pH-stat
   Description: Hydrolysis of the triglyceride results in fatty acids, which resonate between the protonated and unprotonated forms. The pH-stat senses the release of the proton from the protonated form.

Polyphenol Oxidase (EC 1.10.3.1, 1.10.3.2, and 1.14.18.1)

Reaction: Oxidization of a number of mono and polyphenoloic compounds

1. Reference: Boyer, R. F. 1977. Journal of Chemical Education 54:585
   Method: Spectrophotometric (colorimetric).
   Description: The substrate (dihydroxyphenylalanine, Dopa) is oxidized to o-quinone. The reaction is followed by recording the increase in o-quinone absorbance at 475 nm.
2. Reference: Bernier, A. M., and Howes, N. K. 1994. Journal of Cereal Science 19(2):157
   Method: Spectrophotometric (colorimetric).
   Description: Wheat kernels are incubated in tyrosine solution. The absorbance of colored pigments formed is measured in microtitre plate.
3. Reference: Lamkin, W. M., Miller, B. S., Nelson, S. W., Traylor, D. D., and Lee, M. S. 1981. Cereal Chemistry 58:27
   Method: Specific-ion electrode (oxygen)
   Description: Catecholase activity is measured by following the consumption of oxygen as the reaction proceeds.

Alpha-Amylase (3.2.1.1)

Reaction: Endohydrolysis of starch, producing low molecular weight dextrins

1. Reference: American Association of Cereal Chemists. 1995. *Approved Methods of the AACC*, 9th ed. Method 56-81B, Falling Number Determination. The Association, St. Paul, MN.
   Method: Viscometric
   Description: A flour-buffer slurry is heated to boiling in a water-jacketed heater. Enzyme activity is measured as the time required for a plunger to fall through the hot starch paste.
2. Reference: American Association of Cereal Chemists. 1995. *Approved Methods of the AACC*, 9th ed. Method 22-10, Diastatic Activity of Flour, with the Amylograph. The Association, St. Paul, MN.
   Method: Viscometric
   Description: A flour-buffer slurry is heated at a constant rate. As the starch gelatinizes, the viscosity of the hot starch paste increases. Enzyme activity is measured as the decrease in optimal viscosity resulting from hydrolysis of starch.
3. Reference: McCleary, B. V., and Sheehan, H. 1987. Journal of Cereal Science 6(3):237
   Method: Spectrophotometric (colorimetric)
   Description: The substrate is a blocked *p*-nitrophenol maltoheptaoside, which is hydrolyzed by α-amylase (but not by β-amylase). The smaller *p*-nitrophenyl maltosaccharides are then hydrolyzed to glucose and *p*-nitrophenol with an α-glucosidase. The *p*-nitrophenol released is measured at 410 nm. The assay is available from Megazyme International (Bray, County Wicklow, Ireland).
4. Reference: Mathewson, P. R., and Pomeranz, Y. 1979. Journal of the Association of Official Analytical Chemists 62(1):198
   Method: Spectrophotometric (colorimetric)
   Description: A blue-dyed starch is used as substrate for the enzymatic reaction. α-Amylase hydrolysis results in the release of soluble colored dextrins, which are measured at 620 nm.
5. Reference: Ohnishi, M., Iwata, K., and Hiromi, K. 1990. Starch/Staerke 42(3):103
   Method: Spectrophotometric
   Description: The maltohexaose substrate is hydrolyzed by α-amylase. Saccharides are then enzymatically reduced to glucose, which is determined by ultraviolet absorbance. A test kit is available from Boehringer Mannheim Co.

Beta-Amylase (EC 3.2.1.2)

Reaction: Exohydrolysis of starch, resulting in release of maltose

1. Reference: McCleary, B. V., and Codd, R. 1989. Journal of Cereal Science 9(1):17
   Method: Spectrophotometric (colorimetric)
   Description: The method uses p-nitrophenol maltopentaose as substrate in the presence of α-glucosidase. β-Amylase rapidly hydrolyzes the substrate to maltose and p-nitrophenol maltotriose. The maltotriose is hydrolyzed by α-glucosidase to glucose and p-nitrophenol, which is determined at 410 nm. The assay is available from Megazyme.

Cellulase (EC 3.2.1.4)

Reaction: Endohydrolysis of cellulose, releasing glucose

1. Reference: Huang, J. S., and Tang, J. 1976. Analytical Biochemistry 73:369
   Method: Spectrophotometric, fluorometric
   Description: Aminoethyl-cellulose is chemically modified with either trinitrobenzenesulfonic acid (TNBS) or fluorescamine, resulting in a colorimetric or fluorometric substrate, respectively. Hydrolysis of the modified substrate releases a labeled glucose, which is then measured with a spectrophotometer or fluorometer.
2. Reference: Megazyme International (Bray, County Wicklow, Ireland), product literature
   Method: Spectrophotometric (colorimetric)
   Description: Dyed cellulose substrate is hydrolyzed by cellulase, releasing soluble colored products.

Beta-Glucanase (EC 3.2.1.6)

Reaction: Hydrolysis of beta-glucan (both endo- and exo-), resulting in release of glucose and/or saccharides of varying sizes

1. Reference: Megazyme International (Bray, County Wicklow, Ireland), product literature
   Method: Spectrophotometric (colorimetric)
   Description: A dyed β-glucan substrate is hydrolyzed, releasing soluble colored product.

Xylanase (EC 3.2.1.32)

Reaction: Hydrolysis of 1,3- and 1,4-α-D linkages in xylan polymers (pentosan)

1. Reference: Megazyme International (Bray, County Wicklow, Ireland), product literature
   Method: Spectrophotometric (colorimetric)
   Description: A dyed substrate (Xylazyme) is hydrolyzed by xylanase, releasing soluble colored products.

Pullulanase (EC 3.2.1.41)

Reaction: Hydrolysis of the 1,6-α-D-glucosidic bond in polysaccharides such as amylopectin and pullulan

1. Reference: Megazyme International (Bray, County Wicklow, Ireland), product literature
   Method: Spectrophotometric (colorimetric)
   Description: Dyed pullulan is hydrolyzed by pullulanase, releasing soluble colored product.

Pectinesterase (EC 3.1.1.11)

Reaction: Removal of the methyl ester from pectin, resulting in the formation of pectic acid

1. Reference: Hagerman, A. E., and Austin, P. J. 1986. Journal of Agricultural and Food Chemistry 34:440
   Method: Spectrophotometric (colorimetric)
   Description: The formation of pectic acid lowers the pH of an unbuffered solution of pectin in water. The drop in pH can be monitored using a pH-sensitive indicator dye. The reaction is followed by measuring the decrease in absorbance at 620 nm.
2. Reference: Vilarino, C., delGiorgio, J. F., Hours, R. A., and Cascone, O. 1993. Lebensmittel Wissenschaft und Technologie 26(2):107
   Method: Spectrophotometric (colorimetric)
   Description: Formation of pectic acid results in a pH change in the solution. Bromocresol green and Congo red serve as indicator dyes, which allow the reaction to be followed by the color change of the dyes.

Endopolygalacturonase (EC 3.2.1.15)

Reaction: Random hydrolysis of 1,4-α-D-galacturonic acid residues in pectate and other galacturonans

1. Reference: Lopez, P., de la Fuente, J. L., and Burgos, J. 1994. Analytical Biochemistry 220(2):342
   Method: Viscometric
   Description: The substrate solution of polygalacturonate is continuously measured using a rotational viscometer. As endohydrolysis proceeds, the viscosity of the substrate solution decreases.

General Assay for Proteolytic Enzymes

Reaction: Hydrolysis of proteins to amino acids and peptides

1. Reference: American Association of Cereal Chemists. 1995. *Approved Methods of the AACC*, 9th ed. Method 22-62, Proteolytic Activity—Spectrophotometric Method. The Association, St. Paul, MN
   Method: Spectrophotometric, a modification of the Ayre-Anderson hemoglobin method
   Description: Hemoglobin is used as substrate for the enzyme. After an appropriate time, trichloroacetic acid is added to the reaction solution to precipitate protein and peptides. The reaction mixture is filtered, and enzyme activity measured by absorbance at 275 nm.

2. Reference: Adlers-Nissen, J. 1979. Journal of Agricultural and Food Chemistry 27:1256
   Method: Spectrophotometric (colorimetric)
   Description: The free amino groups formed by peptide bond hydrolysis are labeled with trinitrobenzenesulfonic acid, resulting in an adduct that absorbs light at 340 nm.
3. Reference: Mathewson, P. R., Seabourn, B. W., and Pomeranz, Y. 1988. Journal of Cereal Science 8:69
   Method: Fluorometric/spectrophotometric
   Description: As protein is hydrolyzed, newly formed amino groups are labeled with o-phthaldialdehyde in the presence of ethanethiol. This new adduct is highly fluorescent (emission 450 nm) and also absorbs light at 340 nm.
4. Reference: Shotten, D. M. 1970. Methods in Enzymology 19:113
   Method: Spectrophotometric (colorimetric)
   Description: Elastin dyed with Congo red is used as a substrate for elastase. The enzyme releases small, soluble, colored peptides, resulting in a colored solution, the absorbance of which is determined at 495 nm. The basic methodology could be applied to many proteases.
5. Reference: Fukal, L., Kasafirek, E., Strejcek, F., and Kas, J. 1987. Journal of Food Biochemistry 11(2):99
   Method: Spectrophotometric (colorimetric)
   Description: Substrates N-acetyl-Leu-Leu-Gly-p-nitroanilide and N-succinyl-Ala-Ala-(S-benzyl)-Cys-p-nitroanilide are very sensitive to attack by papain. The released p-nitroaniline is measured at about 410 nm.
6. Reference: Boudjellab, N., Rolet-Repecaud, O., and Collin, J. C. 1994. Journal of Dairy Research 61(1):101
   Method: Enzyme-linked immunosorbent assay (ELISA)
   Description: ELISA technology is used to determine the amount and distribution of residual chymosin.

Urease (EC 3.5.1.5)

Reaction: Hydrolysis of urea to carbon dioxide and ammonia

1. Reference: Brocklehurst, K. 1992. Page 210 in: *Enzyme Assays: A Practical Approach*. R. Eisenthal and M. J. Danson, editors. Oxford University Press.
   Method: pH-stat
   Description: The carbon dioxide formed in the reaction forms carbonic acid in water. The pH-stat follows the evolution of CO_2 as carbonic acid, continually titrating with the base to maintain a constant pH. An ammonia-specific ion electrode should also be useful for assay of this enzyme.
2. Reference: ALTECA, Ltd. (Manhattan, KS), product literature.
   Method: spectrophotometric (colorimetric)
   Description: The reaction is proprietary. After reaction, the resulting color is visually compared with a six-point color scale to determine results.

Glossary

Acid protease—A type of proteolytic enzyme that has carboxylic acid functions (often from aspartic acid residues) at the active site and that functions in an acidic pH range (2–4). Examples: pepsin, chymosin, cathepsin D.

Active site—The region on the surface of an enzyme where catalytic activity occurs.

Alkaline (serine) protease—A type of proteolytic enzyme, typically having a serine and a histidine residue in its active site and having activity in the alkaline pH range (about pH 7–11). Examples: trypsin, chymotrypsin, elastase.

Amino acids—A group of organic compounds having the general formula $NH_4C_2O_2R$. Structurally, each is a carboxylic acid with an amino group attached to the α-carbon atom. R represents a functional group peculiar to each amino acid.

Aminopeptidase—Exoprotease hydrolyzing the amino acid at the amino end of the protein.

Amylase—Enzyme that hydrolyzes starch to dextrins and/or sugars.

Amyloglucosidase (glucoamylase)—An enzyme that hydrolyzes both 1,4-α-D- and 1,6-α-D-glucosidic bonds in carbohydrataes, although at different rates.

Amylopectin—The type of starch molecule that has branches.

Amylose—The type of starch molecule that occurs as a linear coil with no branching.

β-Elimination—A type of chemical reaction, not involving water, in which a specific group such as a hydrogen or hydroxyl is removed, resulting in the formation of a double bond.

β-Glucanase—An endoglucanase enzyme that hydrolyzes the 1,3- and 1,4-β-D-glucosidic bonds in β-glucan.

β-Limit dextrin—The oligosaccharides resulting from the limited action of β-amylase (i.e., it cannot hydrolyze the 1,6-α-D-glucosidic linkage) in amylopectin.

Bromelain—Enzyme similar to papain derived from pineapple.

Carboxypeptidase—Exoprotease hydrolyzing the amino acid on the carboxy end of a protein.

Catalase—Enzyme that breaks down hydrogen peroxide to oxygen and water.

Cellulase—An enzyme that hydrolyzes cellulose, a polymer of β-1,4-linked glucose molecules, which is a common component of plant cell walls.

Chymosin—An acid protease, also called rennin, derived from the digestive tract of ruminants, used in the dairy industry.

cis—Isomer in which both hydrogen atoms at a double bond are on the same side.

Cloud loss—Precipitation of the cloudy portion of citrus fruit juice.

Colorimetric—Describing an analytical method in which the interaction of the analyte with light in the visible range is measured, usually as the absorbance of that light.

Covalent bond—A chemical bond formed when two atoms share an electron.

Debranching enzyme—Enzyme capable of hydrolyzing the branch points (e.g., the 1,6-α-D-glucosidic linkages) found in carbohydrate molecules such as amylopectin.

Denaturation—The loss of normal spatial arrangements in a polypeptide chain.

Depectinization—The breakdown of the pectin polymer with enzymes.

Depolymerization—Breakdown of a polymer.

Dextrins—Low molecular weight polymers of glucose (usually less than 10 glucose units in length) resulting from the enzymatic degradation of starch.

Dextrose equivalent—A measure, based on the reducing power of glucose, of the percentage of glucosidic bonds in a starch polymer that have been hydrolyzed.

Diastase—An early name for the α-amylase enzyme.

Disulfide bond—A sulfur-sulfur bond formed between two cysteine amino acid residues, which helps to stabilize the folded structure of proteins.

Endoenzyme—An enzyme capable of splitting bonds anywhere along a polymer chain.

Endogenous enzymes—Enzymes occurring naturally in the substance (e.g., in a plant).

Energy of activation—The minimum energy required to convert a normal reactant molecule into a reactive intermediate.

Enzymatic browning—Darkening of food products resulting from the action of polyphenol oxidases.

Essential oils—The oil fraction in plants (e.g., in citrus fruit peel) having strong flavor and aroma characteristics.

Esterase—An enzyme capable of hydrolyzing an ester linkage, resulting in an acid and an alcohol.

Exoenzyme—An enzyme capable of splitting only terminal bonds in a polymer chain.

Exogenous enzymes—Enzymes from another source added to a substance (e.g., microbial enzymes added to plant substances).

Glucose oxidase—Enzyme that hydrolyzes glucose to gluconic acid and hydrogen peroxide.

Good Manufacturing Practices—Defined practices for ensuring safety and accountability in manufacturing processes.

GRAS—A Food and Drug Administration regulatory status meaning Generally Recognized As Safe.

Hemicellulase—Any of several enzymes, including pentosanase, glucanase, and cellulase, that hydrolyze nonstarch polysaccharides.

Hydrogen bonds—Relatively weak interactions between a hydrogen and an electronegative atom such as oxygen or nitrogen.

Hydrolysis—The breaking of a chemical bond(s) through a water-mediated decomposition mechanism.

Immobilized enzymes—Enzymes that are chemically bound to an insoluble support material in such a way as to maintain the normal activity of the enzyme.

Induced-fit model—A theory to explain enzyme specificity in which a flexible active site is induced, by a substrate, to alter its conformation to an orientation properly fitting the substrate's geometry.

Invertase—Enzyme catalyzing conversion of sucrose to component sugars, glucose and fructose.

Isomerization—The interconversion of different isomeric forms of the same compound (such as the enzymatic conversion of glucose to its isomer, fructose).

Lipase—An enzyme that hydrolyzes a triacylglycerol, forming free fatty acids.

Lipoxygenase—An enzyme that catalyzes the oxidation of unsaturated fatty acids containing a *cis-cis* penta-1,4-diene unit to the corresponding monohydroperoxide.

Liquefaction—The reduction in viscosity of a gelatinized starch slurry, usually accomplished using a heat-stable α-amylase of bacterial origin.

Lock and key model—A theory to explain enzyme specificity in which a substantially rigid active site is likened to a lock and the substrate to a key that fits the lock.

Lyase—An enzyme that removes specific groups from their substrates, usually by elimination rather than hydrolysis, leaving a double bond.

Lysis—Breakdown of the cell wall of a microorganism, which releases its contents.

Mash—The mixture of malted barley and water constituting the first step in brewing beer.

Metalloprotease—A type of proteolytic enzyme having a metal atom, usually Zn, at the active site and having a pH range that centers about pH 7.0. Examples: carboxypeptidase A and B, aminopeptidases, and dipeptidases.

Mobility number—A number corresponding to the reciprocal of the peak viscosity number, that is, the value represented by (1/peak viscosity number).

Monoterpenes—Organic flavor compounds in wine.

Papain—Proteolytic enzyme with broad specificity, derived from papaya.

Peak viscosity number—A number representing the highest value of viscosity attained during an amylograph test.

Pectinase—Enzyme that degrades pectic substances.

Pepsin—An acid protease, usually derived from the bovine or porcine digestive tract and useful in the dairy industry.

Pitching—Adding the yeast to the wort, initiating fermentation.

Polyphenol oxidases—A group of oxidative enzymes acting on phenolic substrates.

Primary structure—The sequence of amino acids making up a protein, from amino terminus to carboxy terminus.

Protease (proteinase)—Enzyme that hydrolyzes proteins to peptides and/or amino acids.

Proteins—Polymers composed of amino acids. Enzymes are proteins.

Pullulanase—A debranching enzyme capable of hydrolyzing the 1,6-α-D-glucosidic linkages in carbohydrates such as pullulan and amylopectin.

Q_{10} rule—Rule stating that a chemical reaction doubles in rate for every 10 degree C increase in temperature.

Quaternary structure—The overall spatial structure of a protein containing more than one polypeptide chain.

Renaturation—The reestablishment of the normal spatial arrangements in a polypeptide chain.

Rennin—An acid protease originally extracted from ruminant stomachs, used chiefly in cheese processing. A similar enzyme, **rennet**, derived from bacterial sources, is usually used for commercial-scale production.

Rheology—The properties of a material that relate to its deformation and flow characteristics.

Saccharification—Further enzymatic hydrolysis of a liquefied starch slurry to produce sugars, accomplished using glucoamylase and/or pullulanase enzymes.

Secondary structure—Protein configuration stabilized by hydrogen bonds not involving side chains; characterized as an α-helix or a β-pleated sheet.

Specification—Written requirements for formulation, packaging, shipment, and any other relevant characteristics of an ingredient.

Spectrophotometric—A type of analytical method that measures change in the interaction of a compound with light.

Sprouting (of wheat)—A condition in which high moisture promotes premature growth of the kernel while in the plant head. This leads to increased enzymatic activity in flour milled from such kernels, which may be detrimental to certain baked products.

Superoxide dismutase—An enzyme that catalyzes a reaction between a superoxide anion (O_2^-) and hydrogen, producing hydrogen peroxide and molecular oxygen (O_2).

Tertiary structure—Overall three-dimensional folding of a protein chain.

Thiol protease—A type of proteolytic enzyme having a cysteine residue in the active site, operative over a wide pH range but typically between pH 4.5 and 9.5 with miximum around pH 6–7.5. Examples: papain, bromelain, ficin.

trans—Isomer in which the hydrogen atoms at a double bond are on opposite sides.

Trypsin—Proteolytic enzyme having limited specificity, derived from animal gut.

Ultrafiltration—A process that uses a semipermeable membrane to separate fractions based on molecular size.

Viscometric—A type of analytical method that measures the change in flow properties of a liquid.

Wort—The liquid remaining after filtering the finished mash.

Xylanase—An endopentosanase enzyme that hydrolyzes the β-1,4 xylosidic bonds in the xylose polymer backbone of which pentosans are composed.

Index

Acid proteases, 33, 55, 77, 78
Actinoplanes missouriensis, 93
Activity, of enzymes
  analysis, 41–46, 97–101
  chemical basis, 4–6
  in dough, effects, 53
  maintenance in storage, 19–20
Additives, enzymes as, 16, 18
Adjuncts, in brewing, 59–60
Alcohol content, in beer, 62, 63, 64, 70
Alkaline phosphatase, 79, 98
Alkaline proteases. See Serine proteases
Allergic reactions, 23–24
Amino acids
  binding of substrate, 7, 8
  structure, 1–2
Aminopeptidases, 31
α-Amylase, 91
  action on starch, 26
  assays, 99
  in brewing, 59, 60, 61, 62
  EC number, 12
  sources, 49–50
  in sugar production, 73, 74
  tests for, 43, 44
β-Amylase, 27, 49, 91, 100
  in brewing, 59, 61
Amylases, 26, 39
  and baked product characteristics, 53, 54, 55
  sources, 49–50
Amyloglucosidase, 27
Amylograph, 44
Amylopectin, 26
Amylose, 26
Anabolic action, 87
Analysis, of enzyme activity
  assay kits, 45
  assay methods, 42–46
  comparison of results, 45–46
  principles, 41–42
  specific techniques, 97–101
Animals, enzyme sources, 95
Antioxidants, 38, 63
Arabinoxylans, 66, 68
Arthrobacter globiformis, 93

Arylesterases, 36
Aspergillus spp., 76, 77
  meleus, 33, 93
  niger, 13, 15, 33, 64, 93
  oryzae, 13, 94
Ayre-Anderson assay method, 43, 45–46

Bacillus spp., 33, 93
  licheniformis, 74, 93
  subtilis, 13, 26, 93, 93
Bacteria as enzyme sources, 13, 93
Bacterial amylases, 49, 50, 54, 73, 74
Bacterial proteases, 77
Barley malt, 49, 59–60
Beer
  low-calorie, 63–64
  production, 60–64, 70–71
  regulation of enzymes for, 18
Bleaching, by lipoxygenase, 37, 52
Bourbon, 65
Bread
  characteristics, 53–54, 56
  production, 50–53
Brewing, 60–64
Bromelain, 91

Carbohydrases, 25–31
  tests for, 43
Carboxyl esterhydrolase, 36
Carboxyl proteinase, 91
Carboxylesterases, 36
Carboxypeptidases, 31, 33, 91
Carriers, in powdered enzymes, 14, 19
Casein, 76, 77
Catalase, 39, 91, 97
Cellulases, 27–28, 39, 67, 68, 91, 100
Cellulose, 27, 68
Cereal amylases, 50, 60
Cheese
  flavor, 36, 37
  production, 76–77
  soft, 77
Choosing enzymes, 81–87
Chymosin, 33, 76, 91
Citrus fruit juice, processing, 69, 72. See also Noncitrus fruit juice

Cleaning work areas, 21, 24
Cloudiness, in beverages, 68, 69, 70, 71, 72
Code of Federal Regulations, 17
Color, enzyme effects on, 38, 54, 56, 67, 71
Colorimetric tests, 43, 79
Commercial enzymes
  production, 13–15
  and purity, 23, 42
  as supplements, *see* Exogenous enzymes
Commercial sweeteners, 73–75
Cookies, 52, 55–56
Crackers
  dough behavior, 52, 55–56
  product characteristics, 56, 57

Dairy products, 76–77
Debranching enzymes, 27
Denaturation, 10
"Designer" fats and proteins, 87
Diacetyl, in beer, 63
Disporotrichum dimorphosporum, 94
Disulfide isomerases, 86
DNA, transfer, 14, 15
Dough, bread
  consistency, 52
  effect of enzymes on, 50–53, 56, 86
  oxidation, 52–53
"Dustless" enzymes, 24
Dye, linked to substrate, 43

Embden-Meyerhoff-Parnas pathway, 62
Endoenzymes, 25
Endogenous enzymes, 59, 69, 77–78
Endopolygalacturonase, 91, 101
Endothia parasitica, 94
Energy of activation, 4, 5
Enzymatic browning, 37–38
Enzyme Commission (EC) number, 11–12
Enzyme-linked immunosorbent assay, 45
Enzyme-modified cheeses, 77

INDEX

Enzymes
　appropriate use, 82–83
　kinetics, 5–6
　nomenclature, 11–12
　structure, 1–4
Equipment, protective, 24
Esterases, 35–36, 76, 77
Ethanol production, 64–66
Exoenzymes, 25
Exogenous enzymes
　in baking, 49, 50, 51
　in beverage production, 60, 61, 62, 66, 67, 68, 69
　for protein modification, 77, 78
Exopolygalacturonase, 91
Expiration date, 20

Fats
　in doughs, effects, 53, 55
　enzymes affecting, 34–37
Fatty acids, 34–35
FDA. *See* U.S. Food and Drug Administration
Fermentation
　of beer and wine, 62, 67, 70
　and discovery of enzymes, 13
　microbial, 13–14
Ficin, 91
Filtration problems, 29, 61, 66, 67, 68, 70, 71
Firming, of fruits and vegetables, 69
Flavor, enzyme effects on
　in baking, 54
　in beverages, 67, 71
　in cheeses, 36, 37, 76, 77
Flavoring ingredients, 77
Fluorescence, for enzyme assay, 44
Foam, in beverages, 63, 67, 70, 71
Food Additive Amendments of 1958, 17
Food-grade enzymes, 16
　marketing regulations, 17–18
Fruit
　citrus, 69
　color, 39
　dried, 68
　juice production, 29, 67–69
　noncitrus, 67–68
　ripening, 39
Fungal amylases, 49, 50
Fungal proteases, 77, 78
Fungi, as enzyme sources, 14, 76, 93–94

Generally recognized as safe (GRAS), 15, 17, 18
Genetic engineering, 15, 39
β-Glucan, 28, 29, 61, 66
β-Glucanase, 28–29, 91, 100

and filtration of beverages, 61, 62, 67
Glucoamylase, 64, 73, 74, 91. *See also* Amyloglucosidase
Glucose isomerase, 14, 75, 91
Glucose oxidase, 39, 44, 91, 97
Gluten, 51–52, 53, 54, 55, 77, 86
Good Manufacturing Practices, 16
Grapes, high in pectin, 66–67

Handling enzymes, 20–24
Haze, in beverages, 63, 67, 68, 70, 71
High-fructose corn syrup, 74, 75
Humicola insolens, 94
Hydrolases, 12, 91
Hydrolytic rancidity, 37
Hydrolyzed vegetable protein, 77–78
Hydroperoxide, 36

Immobilized enzymes, 75, 87
Indicators, processing, 78–79
Induced-fit model, 8
Ingredients
　in commercial enzymes, 19
　enzymes as, 18
International Union of Biochemistry, 11
International Union of Pure and Applied Chemistry, 11
Invertase, 4, 91
Irritation, from enzyme use, 23
Isomerases, 12, 91
Isomerization, 74–75

Kinetics
　chemical, 4
　enzyme, 5–6
Klebsiella spp., 95
　planticola, 93
Kluyveromyces spp., 95
　fragillis, 14

Leucine aminopeptidase, 91
Level of use, 83, 85
Ligases, 12,
β-Limit dextrin, 27
Lipase, 35, 76, 77, 91, 98
Lipoxygenase, 9, 36–37, 91
　assay, 98
　bleaching by, 52
Liquid enzymes, packaging, 21, 86
Liquifaction of starch, 74
Liquors, 65
Loaf volume, 54, 56
Lock and key model, 7–8
Low-fat/no-fat crackers, 55
Low-moisture baked products, 54–55
Lyases, 12, 31, 91

Maillard reaction, 33, 37, 39, 54
Malting, 59
Mashing, 60–61
Meat and poultry, enzyme use in, 18
Metalloproteases, 33, 91
Methylesterase, 69
Microbes, as sources of enzymes, 13, 14–15
Microbial enzymes, 60, 62. *See also* bacterial and fungal amylases and proteases
Mixing of enzyme solutions, 22, 23
Mixtures of enzymes, 22–23, 42, 83
Mobility number, 45
Moisture level, in storage, 19
Molds, in cheesemaking, 77
Monoterpenes, 67
Mortierella vinaceae, 94
Mucor spp., 76, 94

Neutral proteases, 55, 77. *See also* Metalloproteases
Nomenclature of enzymes, 11–12
Noncitrus fruit juice, processing, 67–68, 71
Northrup assay, 45
Nutritional availability, improvement, 78
Nutritional supplements, 78

Oils
　in doughs, effects, 53
　enzymes affecting, 34–37
　essential, in citrus, 69, 72
Oxidation, 37
Oxidizing agents, replacement, 52
Oxidoreductases, 12, 91

Packaging, 19, 21
Papain, 32, 33, 91
Pasteurization, test for, 79
Peak viscosity number, 45
Pectase. *See* Pectinesterase
Pectate lyase, 91
Pectic acid, 29, 30, 69
Pectic acid lyase, 31
Pectin, 29, 30, 66, 69
Pectin demethoxylase. *See* Pectinesterase
Pectin lyase, 31, 68, 91
Pectin methylesterase. *See* Pectinesterase
Pectinases, 29–31, 39, 66, 67, 68, 69
Pectinesterase, 30, 68, 91, 101
Penicillium spp., 94
Pentosans, 28, 52, 66
Pentosanases, 28, 52
　and baked product characteristics, 53, 55
Pepsin, 9, 33

Peroxidase, 79, 91, 97
pH
  in brewing, 60
  for enzyme assay, 44
  and enzyme reaction rate, 8–10
  in fruit juice processing, 69
  optimum levels for enzymes, 20, 33, 35, 50, 62
  in soda cracker production, 55, 84
Phenol oxidases, 91
pH-stat, 44
Phytase, 15, 91
Pizza crusts, effects of enzymes on, 55, 57
Plants, enzyme sources, 95
Plastein reaction, 87
Polygalacturonases, 30, 69
Polyphenol oxidases, 38, 99
Powdered enzymes
  mixing, 21, 22
  packaging, 19
  production, 14
Problems
  troubleshooting, 56–57, 70–72
  using enzymes to solve, 81–87
Processing aids, enzymes as, 16, 18
Production of enzymes
  commercial, 13–15
  regulations, 15, 17–18
Proteases, 31–34, 39
  assays, 101–102
  with bacterial ferments, 55
  and baked product characteristics, 53, 54, 55
  in brewing, 60, 61, 63, 67
  in cheesemaking, 76, 77
  microbial, 60
  for protein modificaton, 77, 78
  specificity, 22, 31–32
  supplementation with exogenous, 51–52, 61, 62, 67, 69, 77, 78
Proteinases. See Proteases
Proteins
  functionality, 78
  modified, 77–78
  soy, 77, 78
  structure, 1–4, 9
  whey, 78
Proteolytic activity
  tests for, 43, 44–45
Proteolytic enzymes. See Proteases
Protopectin, 29, 39, 66
Pseudosubstrate, 8
Pullulanase, 27, 74, 91, 101

Q-10 rule, 10
Quality control program, need for, 19, 60, 62, 84

Rancidity, 35, 37, 79
Reaction order, 6
Reaction rate
  choosing appropriate, 34
  factors affecting, 11, 12
Renaturation, 10
Rennet, 76
Rennin, 76, 91. See also Chymosin
Rhizopus spp., 76, 95
Ripening, of fruit, 39
Rum, 65

Saccharification, 74
Saccharomyces spp., 59
  cerevisae, 14, 95
Safety precautions, 23–24
Serine proteases, 33, 91
Shelf life, of bread, 54
Side chains, of amino acids, 2, 6, 9, 89–90
SKB method for α-amylase asssay, 43, 45
Sources of enzymes, 13–15, 93–95
Soybeans
  cooking, 79
  protein, 77, 78
Specifications for enzymes, 47, 83
Specificity, of enzymes
  of carbohydrases, 27
  and flavor, 77
  mechanism, 7–8
  of proteases, 22, 31–32, 34
  of rennin (chymosin), 76, 86
Spectrophotometric tests, 42, 43
Standardization, 83
Starch
  conversion to sugar, 73–75
  effects of enzymes on, 25–27
  in fruit juice, 67
  gelatinization, 51, 54, 70
  hydrolysis, 50–51, 54
Storage, of enzymes, 16, 19–20
Streptomyces spp., 93
  griseus, 13
Structure of proteins, 1–4, 34
Substrate
  concentration, 6, 11, 42
  interaction with enzyme, 25
  orientation by enzyme, 6–8
  pH level, 8–9

  temperature, 10
Sugar production with enzymes, 73–75
Superoxide dismutase, 38–39
Sweeteners, commercial, 73–75
Syrups, 74, 75

Temperature
  and enzyme activity, 50, 55, 61, 85
  and enzyme reaction rate, 10
  and microbial enzymes, 60
  in storage, 16, 19
  of substrate, 20, 22
Texture, of bread, 53–54
Thiol proteases, 33
Timing of enzyme use, 83–84, 85–86
Transferase, 12
Trichoderma spp., 95
Troubleshooting, 56–57, 70–72
Trypsin, 32, 33, 34
Trypsin inhibitors, 79

U.S. Bureau of Alcohol, Tobacco and Firearms, 17, 18
U.S. Department of Agriculture, 17, 18
U.S. Food and Drug Administration, 15, 17, 18, 63
Urease, 79, 102

Vegetables, 39, 79
Ventilation, need for, 24
Viscometric tests, 42, 44
Viscosity
  and endo- or exoenzymes, 25, 33, 34, 52
  and filtration problems, 29, 61, 66, 67, 68, 70, 71

Water, in dough, 52, 54, 55
Waxy maize starch, 74
Wheat, 49, 50
Whiskey, 65
Wine, enzymes in, 18, 66–67, 71
Wohlgemuth method for α-amylase assay, 43
Wort, enzymes in, 61–62

Xylanase, 28, 67, 91, 100
Xylose isomerase, 75, 91

Yeasts
  in brewing, 59, 62
  as enzyme sources, 14, 93